Advanced Composites World Reference Dictionary

Advanced Composites World Reference Dictionary

John J. Morena

KRIEGER PUBLISHING COMPANY
MALABAR, FLORIDA
1997

Original Edition 1997

Printed and Published by
**KRIEGER PUBLISHING COMPANY
KRIEGER DRIVE
MALABAR, FLORIDA 32950**

Copyright © 1997 by Krieger Publishing Company

All rights reserved. No part of this book may be reproduced in any form or by any means, electronic or mechanical, including information storage and retrieval systems without permission in writing from the publisher.
No liability is assumed with respect to the use of the information contained herein.
Printed in the United States of America.

FROM A DECLARATION OF PRINCIPLES JOINTLY ADOPTED BY A COMMITTEE OF THE AMERICAN BAR ASSOCIATION AND COMMITTEE OF PUBLISHERS:
This publication is designed to provide accurate and authoritative information in regard to the subject matter covered. It is sold with the understanding that the publisher is not engaged in rendering legal, accounting, or other professional service. If legal advice or other expert assistance is required, the services of a competent professional person should be sought.

Library of Congress Cataloging-In-Publication Data

Morena, John J., 1937–
    Advanced composites world reference dictionary / John J. Morena.
       p.   cm.
    ISBN 0-89464-991-4 (alk. paper)
    1. Composite materials—Dictionaries.   2. High technology—
Dictionaries.   I. Title.
TA418.9.C6M66   1997
620.1′18—dc21   96-49072
                                                              CIP

10 9 8 7 6 5 4 3 2

PREFACE

During this decade and the next, there will be an increasing need for industrial change from military, defense, aircraft, aerospace and other vehicle and infrastructure manufacturing to applications in the commercial sector. Technology transfer in Advanced Composites will lead the way to this transition. A basic understanding of the terms, phrases, definitions, and acronyms will be prerequisite.

Advanced Composites World Reference Dictionary is a global reference book. An easily read format will allow the book to be used for all level educational programs as well as informing and guiding managers, designers, engineers, technicians, shop, mold, tool and parts fabrication staff and others.

This book, with its illustrations, is a collectoin of domestic and international words, terms and definitions collected from advanced composites education and industry leaders.

In this regard, I am indebted to many friends and industry technologists including Dr. John Summerscales and his associates from the U.K. and E.C., Diana Votruba for her consistent administrative support throughout the manuscript preparation, Cynthia Murphy; Dr. Jerzy Nowacki, Poland; Dr. J. Bouwma, The Netherlands; and Vickie P. McConnell, Sr. Editor.

ABBREVIATIONS AND ACRONYMS

***NOTE:** The abbreviations and acronyms section contains entries that are not in the text. The section also contains entries for which more than one definition is listed.

ABR	acrylate butadiene rubber	**CA**	cellulose acetate
ABS	acrylonitrile butadiene styrene	**CAB**	cellulose acetate-butyrate
ACM	acrylate rubber	**CAD**	computer aided design
AE	acoustic emission	**CAM**	computer aided manufacturing
AES	auger electron spectroscopy	**CAP**	cellulose acetate-propionate
AEPS	acrylonitrile ethylene propylene styrene	**CARP**	committee on acoustic emission from reinforced plastics
AESL	acoustic emission source location		
AI	amide-imide resins and polymers	**CAT**	computer-aided tomography
AIMS	advanced integrated manufacturing system	**CAT-SCAN**	computer-aided tomography
		CB	core box
AJ	assembly jig	**CC**	carbonaceous heat-shield composites (C-C is carbon-carbon)
AM	assembly model		
AMMA	acrylonitrile methyl methacrylate copolymer	**CCA**	cellular cellulose acetate
		CCRP	carbon or graphite cloth reinforced plastic
ANM	acrylonitrile acrylate rubber	**CF**	checking fixture
APC	aromatic polymer composite	**CFG**	compact-flake-graphic iron
API	addition-reaction polyimide	**CFR**	cresol-formaldehyde resin
APP	atactic polypropylene	**CFRP**	carbon or graphite fiber reinforce plastic
APT	applied potential tomography	**CGT**	computer generated tool
AQL	acceptable quality level	**CHC**	epichlorohydrin-ethylene oxide rubber
ASA	acrylonitrile styrene acrylate terpolymer	**CHR**	epichlorohydrin rubber (*see also* CO)
ATLAS	automated tape lay up system	**CIL**	thermoplastic material flow test
ATR	attenuated total reflection	**CIM**	computer integrated manufacturing
ATT	apply trim template	**CKT**	cloth kit template
AU	acousto ultrasonics	**CL**	chemiluminescence
AULE	arbitrary units of light emission	**CMC**	carboxymethyl cellulose
AUSS	automatic ultrasonic scanning system	**CN**	cellulose nitrate (*see also* NC)
		CNR	carboxynitroso rubber; (tetrafluoroethylene-trifluoronitrosomethane-unsat.monomer terpolymer)
BD	blanking die		
BFD	blank & form die		
BIIR	brominated isobutene-isoprene (butyl)		
BMC	bulk molding compound	**CO**	poly[(chloromethyl)oxirane]; epichlorohydrin rubber (*see also* CHR)
BMI	bismaleimide		
BOF	bonding or boring fixture	**CP**	caul plate
BPHB	bag press hydro block	**CP**	cellulose propionate
BPFD	blank, pierce & form die rubber	**CP**	cross polarization
BR	cis-1,4-butadiene rubber (cis-1,4-polybutadiene)	**CP**	cross ply
		CPE	chlorinated polyethylene
BS	butadiene-styrene copolymer (*see also* SB)	**CPIs**	condensation-reaction polyimides
		CP-MAS	cross polarization with magic angle spinning in NMR
BT	blanking tool		

CR	chloroprene rubber	EPDM	ethylene-propylene rubber (*see also* E/P, EPR)
CS	casein		
CSA	chemical shift anisotropy	EPE	epoxy resin ester
CSM	chopped strand mat	EPM	ethylene-propylene rubber (*see also* E/P, EPR)
CT	computed tomography, or computer tomography		
		EPR	ethylene-propylene rubber (*see also* E/P, EPM)
CT	contour template		
CTA	cellulose triacetate	EPS	expanded polystyrene; polystyrene foam (*see also* XPS)
CTFE	poly(chlorotrifluoroethylene); (*see also* PCTFE)		
		ERM	elastic-reservoir molding
CW	continuous wave	E.S.C.	environmental stress cracking
CVD	chemical vapor deposition	ESD	electrostatic discharge
		ESE	electric screening effectiveness
DDA	dynamic dielectric analysis	ESPI	electronic speckle pattern interferometry
DETA	dielectric thermal analysis		
DF	drill fixture	ESR	electron spin resonance
DFA	diffuse field analysis	ETFE	ethylene-tetrafluoroethylene copolymer
DIB	diiodobutane	EVA,E/VAC	ethylene-vinyl acetate copolymer
DJ	drill jig	EVE	ethylene-vinyl ether copolymer
DMC	dough molding compound	EWGAE	european working group on acoustic emission
DMTA	dynamic mechanical thermal analysis		
DR	diffuse reflectance		
DRD	draw die	FB	form block
DS	directionally solidified	FBT	form block template
DSC	differential scanning calorimetry	FCE	fluorine-containing elastomer
DT	drill tool	FD	form die
DTA	differential thermal analysis	FDI	frequency domain interferometry
		FE	fracto-emission
EAA	ethylene-acrylic acid copolymer	FEP	florinated ethylenepropylene (*see also* PFEP)
EAM	ethylene-vinyl acetate copolymer		
EATF	externally applied thermal field	FF	furan-formaldehyde resins
EC	ethyl cellulose	FFT	fast fourier transform
ECB	ethylene copolymer blends with bitumen	FID	free induction decay
		FMD	foam mold die
ECD	electrostatic charge decay	FOD	foreign-object damage
ECF	envelope check fixture	FP	polycrystalline alumina
ECTFE	ethylene-chlorotrifluorethylen copolymer	FP Fiber	polycrystalline alumina fiber
		FPM	fiber percolation model
EDM	electrical discharge machining	FRAT	fiber reinforced advanced titanium
EE	electron emission in fracto-emission monitoring	FRP	fiber reinforced plastic
		FSI	fluorinated silicone rubber
EEA	ethylene-ethyl acrylate copolymer	FT-NMR	fourier transform nuclear magnetic resonance
EFM	electroform mandrel		
EMA	ethylene-methacrylic acid copolymer or ethylene-maleic anhydride copolymer	FTIR	fourier transform infrared spectroscopy
EMC	elastomeric-molding tooling compound	GBP	generalized weighted back projection
EMF	electromotive force	GF	grinding fixture
EMI	electromagnetic interference	GPC	gel permeation chromatography
EMT	exterior marking template	GR-I	butyl rubber, former US acronym. See IIR PIBI
EP	epoxy resin		
E/P	ethylene-propylene copolymer (*see also* EPM, EPR)	GR-N	nitrile rubber, former US acronym. See NBR

GR-S	styrene-butadiene rubber, former US acronym. *See* PBS, SBR	MAR	magic angle rotation
		MAS	magic angle spinning
		MBS	methyl methacrylate-butadiene-styrene terpolymer
HCCD	honeycomb crush die		
HCT	hole checking template	MC	methyl cellulose
HD	hammer die	MCD	master control drawing
HDPE	high density polyethylene	MDE	microdielectrometry
HEC	hydroxyethylcellulose	MDPE	medium density polyethylene (ea. 0.93–0.94 g/cm^3)
HFD	hydroform die		
HIP	hot isostatic pressing	MF	melamine-formaldehyde resin
HIPS	high impact polystyrene	MG	master gage
HLT	hole locating template	MIR	multiple internal reflection
HM	high-modulus	MM	master model
HME	high-vinyl-modified epoxy	MPF	melamine-phenol-formaldehyde resin
HMM	hollow mold mandrel	MR	magnetoresistance
HMWPE	high molecular weight polyethylene	MRI	magnetic resonance imaging, or medical resonance imaging
HSRTM	high speed resin transfer molding		
		MTT	master tooling template
ICAM	integrated computer-aided manufacturing	MIT	miscellaneous tool
		MVT	moisture vapor transmission
IFAC	integrated-flexible automation center		
IIR	isobutene-isoprene rubber; butyl rubber (*see also* GR-I, PIBI)	N	acoustic emission count (qv)
		NASTRAN	general purpose computer code for finite element analysis laminates
ILC	integrated laminating center	NBR	acrylonitrile-butadiene rubber; nitrile rubber; GR-I
IM	intermediate-modulus and also injection molding		
		NC	nitrocellulose; cellulose nitrate (*see also* CN)
IPN	interpenetrating polymer network		
IPN	interpenetrating network	NC	numerically controlled as a computer
IQI	image quality indicator		
IR	synthetic cis-I,4-polyisoprene	NCR	acrylonitrile-chloroprene rubber
IRS	internal reflection spectroscopy	NCT	numerical control tape
IT	interim tool	NDE	nondestructive evaluation, similar to NDI
		NDI	nondestructive inspection
JDT	jig drill template	NDT	nondestructive testing, similar to NDI
JIT	just in time	NE	neutral emission
		NEPS	small bunches of tangled fibers found in fabrics and yarns
KHD	kirksite hammer die		
		NIE	negative ion emission
LDPE	low density polyethylene	NIOSH	national institute for occupational safety and health
LHS	low-cost high-strength		
LIM	liquid injection molding	NIR	acrylonitrile-isoprene rubber
LJ	locating jig	NMA	nadic mthyl anhydride, accelerator for epoxy resins
LLDPE	linear low density polyethylene		
LLW	leaky lamb waves	NMR	nuclear magnetic resonance
LM	layup mandrel	NQR	nuclear quadrupole resonance
LMC	low-pressure molding compound	NR	natural rubber (cis-I,4-polyisoprene)
LOCAN	location analizer	N/C	numerical control
LST	large space telescope		
LWV	lightweight vehicle	OER	oil extended rubber
		OPR	propylene oxide rubber
MABS	methyl methacrylate-acrylonitrile-butadiene-styrene	OFS	optical fibre sensors
		OT	optical tool

PA	polyamide (e.g. PA 6,6 = polyamide 6,6 = nylon 6,6 in US literature)	PMA	poly methyl acrylate
		PMI	polymethacrylimide
PA	production aid	PMMA	polymethyl methacrylate
PAA	polyacrylic acid	PMMI	polypyromellitimide
PAI	polyamideimide	PMP	poly 4-methyl-1-pentene
PAMS	polyalphamethylstyrene	PO	polypropylene oxide; or polyolefins; or phenoxy resins
PAN	polyacrylonitrile		
PARA	polyarylamide	POM	polyoxymethylene, polyformaldehyde
PAT	pattern	PP	polypropylene
PB	poly 1-butene	P&P	plastic and plaster
PBI	polybenzimidazoles	PFP	plastic-faced plaster
PBMA	poly n-butyl methacrylate	PPC	chlorinated polypropylene
PBR	butadiene-vinyl pyridine copolymer	PPE	poly phenylene ether
PBS	butadiene-styrene copolymer (*see also* GR-S, SBR)	PPMS	poly para-methylstryrene
		PPO	poly phenylene oxide
PBT, PBTP	polybutylene terephthalate	PPOX	poly propylene oxide
PC, PCO	polycarbonate	PPS	poly phenylene sulfide
PCD	polycarbodiimide	PPSU	poly phenylene sulfone
PCTFE	polychlorotrifluoroethylene	PPT	poly propylene terephthalate
PDAP	polydiallyl phthalate	PS	polysulfone
PDMS	polydimethylsiloxane	PS	polystyrene
PE	polyethylene	PSB	polystyrene butadiene rubber (*see also* GR-S, SBR)
PEA	polyethyl acrylate		
PEC	chlorinated polyethylene (*see also* CPE)	PSF, PSO	polysulfone
		PSU	poly phenylene sulfone
PEEK	polyether etherketone	PTFE	poly tetrafluoroethylene
PEI	polyetherimide	P3FE	poly trifluoroethylene
PEO, PEOX	polyethylene oxide	PTMT	poly tetramethylene terephthalate = poly butylene terephthalate (*see also* PBTP)
PEP	ethylene propylene polymer (*see also* E/P, EPR)		
PEPA	polyether polyamide block copolymer	PUR	polyurethane
PES	polyethersulfone	PVA, PVAC	poly vinyl acetate
PET, PETF	poly ethylene terephthalate	PVAL	poly vinyl alcohol (also PVOH)
PF	phenol-formaldehyde resin	PVB	poly vinyl butyral
PFA	perfluoroalkoxy resins	PVC	polyvinyl chloride
PFEP	tetrafluoroethylene-hexafluoropropylene copolymer; FEP	PVCA	vinyl chloride-vinyl acetate copolymer
		PVCC	chlorinated poly vinyl chloride
PFF	preform fixture	PVDC	poly vinylidene chloride
pH	measure of acidity or alkalinity	PVDF	poly vinylidene fluoride
phE	photon emission	PVF	poly vinyl fluoride
PI	polyimide	PVFM	poly vinyl formal
PIB	polyisobutylene	PVI	poly vinyl isobutyl ether
PIBI	isobutene-isoprene copolymer; butyl rubber GR-I, IIR	PVK	poly n-vinylcarbazole
		PVP	poly n-vinylpyrrolidone
PIBO	polyisobutylene oxide	PVT	pulsed video thermography
PIC	pressure-impregnation-carbonization		
PIE	positive ion emission	QA	quality assurance
PIP	synthetic poly-cis-1,4-polyisoprene; (*see also* CPI, IR)	RAPS	radar absorbing primary structure
		RASP	radar absorbing structural panels
PIR	polyisocyanurate	RCF	roller coat fixture
PLT	ply locating template	RF	radio frequency
PM	part model		

R.F.	radio frequency	TJ	trim jig
R.F.P.	radio frequency preheating	TLM	tape-laying machine
RF	router fixture	TMA	thermal mechanical analysis
RFI	radio frequency interference	TMC	thick molding compound
RFR	resorcinol formaldehyde resin	TMLO	tooling master layout
RFW	radio frequency welding	TMR	topical magnetic resonance
RIM	reaction injection molding	TOFD	time of flight diffraction
RRIM	reinforced reaction injection molding	TPE	thermoplastic elastomer
RM	rotational mold	TPI	turns per inch
RMS	root mean square	TPR	1,5-trans-poly (pentenamer)
RP	reinforced plastic	TPU	thermoplastic polyurethane
RPP	reinforced pyrolyzed plastic	TPX	poly (methyl pentene)
RRS	resonance raman scattering	TSA	thermoelastic stress analysis
RTM	resin transfer molding	TSB	tool subbase
RTV	room temperature vulcanizing	T1	longitudinal relaxation time (spin-lattice interaction) in NMR
SAFT	synthetic aperture focusing technique	T2	transverse relaxation time (spin-spin interaction) in NMR
SAM	scanning acoustic microscopy		
SAN	styrene-acrylonitrile copolymer		
SAP	sintered-aluminum powder	UDC	unidirectional composites
SB	styrene-butadiene copolymer	UF	urea-formaldehyde resins
SBR	styrene-butadiene rubber (*see also* GR-S)	UHM	ultrahigh-modulus
SCR	styrene-chloroprene rubber	UHMW	ultrahigh molecular weight abreviation for use with polyethylenes
SDFD	stretch draw form die		
SEA	surface energy analysis		
S-EPDM	sulfonated ethylene-propylene-diene terpolymers	UHMPE	*See* UHMW-PE
		UHMW-PE	ultrahigh molecular weight poly(ethylene) (molecular mass over 3.1×10^6 g/mol)
SF	sample part		
SGTF	stress generated thermal field		
SHIPS	super-high impact polystyrene	UL	underwriters laboratory
SI	silicone resins; poly (dimethylsiloxane)	UT	utility tool
SIR	styrene-isoprene rubber	UV	ultraviolet
SMA	styrene-maleic anhydride copolymer	UVS	ultraviolet stabilizer
SMC	sheet-molding compound		
SMS	styrene-alpha-methylstyrene copolymer		
SPATE	stress pattern analysis by thermal	VARI	vacuum assisted resin injection
SPF/DB	superplastic-forming diffusion bonding	VC/E	vinyl chloride-ethylene copolymer
SPMC	solid polyester molding compound	VC/E/VA	vinyl chloride-ethylene-vinyl acetate copolymer
SRIM	structural reaction injection molding		
SSI	speckle-shearing interferometry	VC/MA	vinyl chloride-methyl acrylate copolymer
STA	simultaneous thermal analysers		
STFB	stretch form block	VC/MMA	vinyl chloride-methyl methacrylate copolymer
SUT	setup template emission		
SWE	stress wave emission	VC/OA	vinyl chloride-octyl acrylate
SWF	stress wave factors	VC/VAC	vinyl chloride-vinylidene chloride
		VDP	van der pauw technique
TA	tooling aid	VFM	vacuum forming mold
TD	trimming die	VHP	vacuum hot pressing
TDI	toluene diisocyanate	VIM	vibrational microlamination
TG	glass transition temperature	VPI	vibration pattern imager
TGA	thermogravimetric analysis		
TFRS	tungsten-fiber-reinforced superalloy	WFS	wet flexural strength

XLPE	cross-linked polyethylene	**XPS**	expandable or expanded polystyrene; (*see also* EPS)
XMC	directionally reinforced molding compound	**Z**	microwave wave impedance

A

A-basis: Mechanical property value, "A" (allowable), is the value above which at least 99% of the population of values is expected to fall, with a confidence of 95%. An A-basis allowable may be used for a single load path structure or one not subjected to a structural test.

A-stage: The period of time, early in a reaction of thermosetting resin system where the material chemical structure is still linear, soluble in certain liquids and is fusible. *See also* B-stage and C-stage.

ABL bottle: Vessel made by filament winding. Used to test internal pressure, mechanical and other properties including quality. Dimensions about 24 inches (610 mm) long and 18 inches (460 mm) in diameter.

ablation: Thermal energy is expended by the sacrificial loss of surface region material. The surface temperature is controlled and the flow of heat is greatly restricted into the material interior. Pyrolysis takes place near the material surface exposed to the heat.

ablative plastic: A material that is being consumed while it absorbs heat through a pyrolysis decomposition process that takes place at or near the exposed surface.

abrasion: The wearing away of a surface by scraping or rubbing.

abrasion resistance: The ability of a material or surface to resist wear. This term is usually used in reference to the ability of a pattern, mold tool or structural composite part to withstand exposure to sand or other abrasive media. If material is a coating, the ability to resist rubbing or impingement upon the surface.

abrasive: Any substance used for abrading, such as grinding, sanding, polishing or sand blasting.

abrasive trimming: Adjusting the size of an item or component by changing its dimensions using an abrasive medium (water jet, grinding equipment) against the surface edge.

absorbed contaminant: A captive contaminant attracted to the surface of a material usually in the form of a vapor, gas, condensate or particulate.

absorption: The penetration into the mass of one substance by another. In a field of radiation energy dissipation occurs. In a liquid, capillary or cellular attraction my occur.

accelerated aging: A method for deterioration and simulation of natural aging artificially hastened and reproduced.

accelerated test: A test procedure in which conditions of accelerated aging are increased to reduce the time required to obtain a deteriorating effect final result.

accelerator: An additive used to speed up a chemical reaction between the resin and catalyst. Reduces curing, hardening or polymerization time of resin systems or vulcanizing elastomers. A promoter.

acceptable quality level (AQL): A satisfactory process average measured by the maximum number of defects per one hundred units.

acceptance test: A test or tests performed upon receipt of goods that determine material conformance to the requirements of a specification, drawing, purchase order or other documents as defined and agreed between the purchaser and supplier.

accumulator: A blow molding equipment reference term that designates an auxiliary ram extruder which is used to provide fast shot or parison delivery. The extruder fills the accumulator cylinder between shot or parison deliveries and is stored until the next shot delivery.

accuracy: The measured or observed value deviation from the actual or accepted reference.

acetal resins: Acetal linkage polymers such as polyoxymethylene.

acid acceptor: An absorbing or combining compound that acts as a stabilizer with acid. May be initially present in minute quantities in a plastic or which may be formed by the decomposition of the resin.

acid-core solder: Wire solder containing acid flux.

1

acid flux: An aqueous solution of an acid and water soluble, organic or inorganic flux.

acid number: The amount or measure in milligrams of potassium hydroxide that is required to neutralize one gram of a substance.

acoustic emission: A nondestructive measure of integrity of a material, as determined by sound amplitude and frequency. These emissions can be associated with defects within the material.

acrylic ester: An ester of acrylic acid, or derivative of acrylic acid, such as, methyl methacrylate.

acrylic plastic: A thermoplastic or thermoset polymer formed by the polymerization of esters of acrylic acid or their derivatives, sometimes referred to as polymethyl methacrylate.

acrylic resin: Synthetic resins prepared from acrylic acid or derivative of acrylic acid.

acrylonitrile: A monomer, sometimes used as a synthetic fiber. Some copolymers with styrene are tougher than polystyrene.

acrylonitrile-butadiene-styrene (ABS): Three thermoplastic polymer blends or copolymers of polystyrene, acrylonitrile copolymer with butadiene-acrylonitrile rubber.

activation: The radiation, thermal, chemical or other process of making a substance or surface more receptive to reaction with another material such as an adhesive, coating or encapsulant.

activator: An additive used as a promoter for curing matrix resin systems. Reduces curing time. *See also* accelerator.

addition polymer: Polymers formed by chain reaction or addition polymerization.

addition polymerization: A chemical reaction in which simple molecules (monomers) are added to each other to form long-chain molecules (polymers). No by-products are formed.

additive: A substance that is added to another substance. Additives such as light stabilizers, fire retardants, toughening agents, etc., usually improve final properties of a material.

adhere: To cause two or more surfaces to be held together by adhesion.

adherend: A body or surface which is held to another body or surface by an adhesive.

adhesion: The state in which two surfaces are held together by chemical or mechanical interfacial forces or interlocking or both.

adhesion failure: The rupture or separation of an adhesive bond such that the separation is at the adhesive-adherend interface.

adhesion, mechanical: An interlocking between surfaces or substrate or part adhesive holds the parts together by interlocking action.

adhesion promoter: A chemical material applied to a substrate to improve adhesion of other materials to the substrate. Also called primer.

adhesion, specific: The adhesion between two or more surfaces held together by valence forces of the same type that result in cohesion.

adhesive: A substance capable of holding two or more materials together by surface attachment. Usually in the forms of a liquid, paste or self-supporting film.

adhesive, cold-setting: An adhesive that reacts and cures at temperatures below 20°C (68°F).

adhesive, contact: Requires contact to instantly bond apparently dry surfaces which are no further apart than about 0.1 mm after contact.

adhesive, dispersion: A two-phase material with one phase suspended in a liquid.

adhesive, foamed: A reduced density bonding material containing gaseous cells and voids, cells dispersed throughout its mass.

adhesive, heat activated: A dry film that becomes tacky and flows by application of heat and or pressure.

adhesive, hot melt: A molten material that forms a bond upon cooling.

adhesive, hot setting: A material that requires temperatures above 100°C (212°F) to react.

adhesive, intermediate temperature setting: Materials (warm setting) as adhesives that react at temperatures from 31 to 99°C (87 to 211°F).

adhesive joint: The area of contact at which two surfaces or substrates are held together.

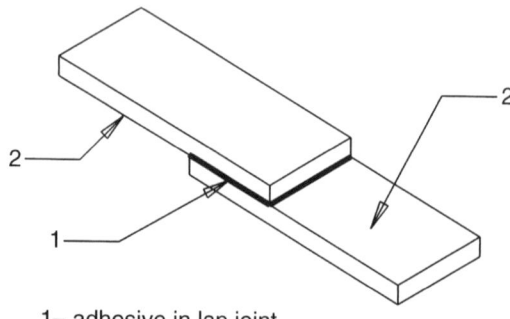

1– adhesive in lap joint
2– surfaces being bonded

Adhesive joint

adhesive, multiple layer: Adhesive film composite, supported or unsupported with different material formulations on each side. Used for bonding dissimilar materials or substrates together.

adhesive, pressure-sensitive: Permanently tacky, solvent free, viscoelastic material that will bond instantly to surfaces (mostly solid) with only slight application pressure.

adhesive, room temperature setting: A material that reacts from 20 to 30°C (68 to 86°F).

adhesive, separate application: A two part bonding system. One part is applied to one adherend and the other part to the other adherend. The two form a joint upon contact.

adhesive, solvent: An adhesive with a volatile organic liquid vehicle.

adhesive, solvent activated: A dry adhesive film that becomes tacky prior to use by following application of a solvent.

adhesive strength: Adhesive to adherend bond strength.

adhesiveness: The adhesion stress property:

$$A = F/S$$

F is force perpendicular to the bend line
S bond surface area.

adiabatic: A process where the heat required to form material or parts is derived from the mechanical action of the process equipment. The contained heat may be removed by cooling to control the process, as in "adiabatic extrusion."

admixture: Homogeneous dispersion and addition of discrete components, before cure.

adsorbate: After adsorption the retained material.

adsorbent: An active surface substance upon which another substance will be adsorbed.

adsorption: The adhesion of the molecules of dissolved substances, gases, or liquids in more or less concentrated form, to the surfaces of liquids or solids with which they are in contact. The concentration of a substance at a surface or interface of another substance.

advanced composites: Highly structural materials with properties comparable to or better than metals. Formed using thermoset or thermoplastic matrix resin systems and high modulus, stiff, reinforcing fibers embedded within the matrix.

after-bake: Heating of fully cured parts to improve mechanical, electrical, thermal and other properties. See post-cure

aging: The change of a material's properties with time, under defined environmental conditions, that leads to improvement or deterioration of properties.

aggregate: Hard, coarse, fragmented material usually used with binders, in tools and flooring applications.

air-assist forming: Thermoforming process is used to form the plastic sheet just prior to final vacuum forming step.

air-bubble void: Entrapment of non interconnected and spherical shaped air pockets within the laminations of reinforcement, an encapsulated or bondline area.

air content: Void or air volume of the pores within connected particles of a substance. In open and closed cell polymeric foam the air is expressed as a percentage of the total volume.

air-dry: Air without water vapor, coating evaporating its solvent vehicle.

air gap: Distance from extrusion coating die opening to the nip, formed by pressure and chill rolls.

air locks: Trapped air causing depressions along a molded part surface and mold interface.

air pollution: Contaminated or unclean air or atmosphere caused by toxic and harmful substances.

air ring: Round mainfold through which cool air or another medium passes and escapes, used to cool plastic blown tubular films during manufacture.

air-slip forming: Thermoforming method which uses a cushion of entrapped air between the plastic sheet and mold surface to keep the mold from contacting the material to be formed until the mold is closed against the vacuum frame. Upon contact the air cushion is released.

air vent: Outlet to prevent entrapment of gases. Usually placed in the surface of a mold or tool.

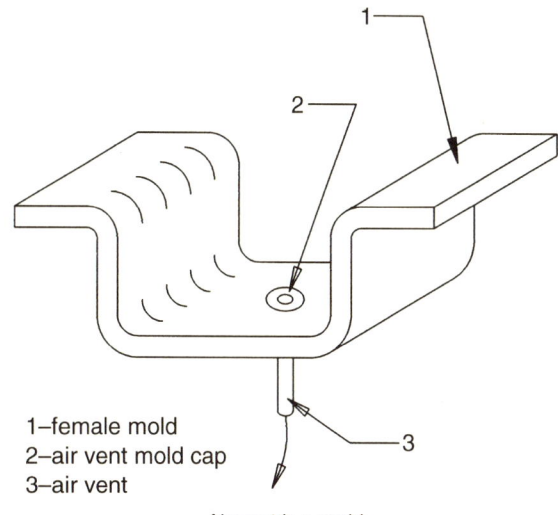

1–female mold
2–air vent mold cap
3–air vent

Air vent in a mold

alclad: Metallic composite sheet material of high purity or corrosion-resistant aluminum bonded to one or both sides of a structural aluminum alloy base.

alkyd plastics: Thermoset materials based on alkyd synthetic resins composed principally of polymeric esters. Recurring ester groups are an integral part of the main polymer chain.

alligatored: Polymeric material surface cracking or degradation with the appearance of the animal skin.

allotropy: The ability of an element or substance to exist in various forms (such as crystal, carbon, graphite).

allowable: A standardized material or part property values used for design (usually mechanical).

alloy: A blend of polymers, copolymers with other polymers. In metals, a material containing more than one element as a metal.

allyl plastics: Thermoset materials based on addition polymerization monomer resins that contain allyl groups (such as diallyl phthalate-DAP).

alternating copolymer: One in which the molecules of each monomer are alternating in the polymer chain repetitively.

alternating stress: Varies between two maximum equal values but of opposite signs. This according to a law determined in terms of the time.

alternating stress amplitude: A test parameter of a dynamic fatigue test where in one cycle there is one-half the algebraic difference between the maximum and minimum stress.

aluminizing: The formation of aluminum or aluminum alloy coatings on a metal or nonmetallic material by diffusion, hot dipping or spraying.

ambient: Surrounding environmental conditions of temperature, pressure or relative humidity such as ambient temperature (temperature in a room).

amide: Organic compound containing $CONH_2$ group, related to organic acids, derivative of ammonia.

amine: Organic base derived from ammonia (NH_3) compound. Forms include aromatic, aliphatic, primary, secondary, tertiary and quarternary.

amine equivalent weight: Amine molecular weight divided by number of molecule active hydrogens.

amine resin: A synthetic resin derived from the reaction of urea, thiourea, melamine or allied compounds with aldehydes, particularly formaldehyde.

Amino: Indicates the presence of an NH_2 group.

amorphous: A material, such as a polymer or resin system, that has no crystallinity (no definite order or pattern to the distribution of the molecules in the material).

amorphous phase: Lacking crystallinity—no crystalline component. Plastics are usually at their amorphous state during processing at temperature.

amorphous plastic: A plastic with no crystalline component, order or pattern to the distribution of the molecules. Random, unstructured molecular configuration.

amylaceous: Relating to the composition and nature of a starch (starchy).

anaerobic adhesive: Adhesive that cures after it has been confined between substrates and is not in the presence of air or oxygen.

anelasticity: Certain materials exhibit properties where strain is a function of both stress and time. A finite time is required to establish equilibrium between stress and strain in both the loading and unloading directions and no permanent deformations may be involved.

angle of contact: If a liquid drop contacts a surface and remains exactly spherical (point contact) then the angle of contact is 180°. Angle between tangent to periphery of point of contact with solid surface. If drop spreads flat, point of contact is zero.

angle of crossing: Angle between a yarn wrap on consecutive passes up and down a spool.

angle of wind: Measured between a warp of yarn on the surface and diametrical plane of the yarn package.

angle-ply: Any combination of filaments on lamina oriented in a specified direction other than 0° that which has been specified as 0° (reference axis).

angle-ply laminate: Equal ply, orthotropic laminate having fibers of adjacent plies oriented at alternating positive and negative (±45°) angles.

angle press: Hydraulic molding press with horizontal and vertical rams. Designed for the production of deep, complex, undercut molded parts.

angle wrap: Fabric tape wrapped around a mandrel at an angle to the centerline.

aniline-formaldehyde resins: High dielectric strength, chemically resistant, thermoplastics of the aminoplastic family. Produced by condensation of formaldehyde and aniline in an acid solution.

anionic polymerization: Polymerization propagated by negatively charged carbon atom or groups containing unshared pair of electrons.

anistropic: Exhibiting different material properties in response to stress applied along axes in different directions.

anisotropic laminate: One in which strength properties are different in various directions.

anisotropy: Varying material properties based upon the orientation or direction of the reference coordinates.

annealing: Stress-relieving process often used on thermoplastic and thermoset nonmetallic materials and parts. To relieve internal and surface stresses without distortion of the shape by holding material at an elevated temperature but below the melting or decomposition point. Use of heat to treat metallic materials and parts results in molecular reorientation, grain structure and property changes.

antifoaming agent: Solution or emulsion, surface tension reducing additive or defoaming agent. Eliminates or breaks surface foam in polymeric coatings or liquids upon deaeration or degassing.

antioxidant: Inhibition or decomposition additives that slows or prevents oxidative and other types of degradation of plastic materials exposed to the atmosphere.

antistatic agent: Chemical and metallic additions which reduce static electricity and when added to the molding material or applied on the surface of the molded object, make it less conducting thereby hindering the fixation of dust.

antisymmetry: In a laminate where component ply segments have unsymmetric angles and sign changes between off-diagonal components.

apparent density: Weight per unit volume of powder or porous material, determined by using a specified loading method.

applicator: Item used to apply polymeric or other liquid materials. Tooling aid such as brush, blade, roller, squeegee, etc.

aramid: A type of highly oriented polyamide material with an aromatic ring structure. Aramid as fibers have excellent high temperature, electrical, high strength and modulus, good fatigue resistance, low coefficient of thermal expansion and are used for producing light weight composites.

arc resistance: Ability for a composite or plastic surface to withstand exposure to an electric voltage. The effects of a high voltage, low current arc should not result in rendering the surface conductive through carbonization by the arc discharge.

areal weight: Mass or weight of one ply or layer of dry fiber fabric, tape or reinforcement.

aromatic: Unsaturated hydrocarbon containing one or more benzene ring structures within the molecule and usually referring to compounds or solvent mixtures.

artificial weathering: Also called artificial aging. This term describes laboratory conditions, usually accelerated and intensified, used for testing plastics, composites and materials in a simulated outdoor exposure environment. Direct water spray, humidity, wind, temperature and ultravoilet energy are some conditions used.

ash content: The solid residue remaining after a reinforcing or inorganic substance has been strongly heated, decomposed or incinerated.

aspect ratio: The ratio of length to diameter of a discontinuous fiber. Also the length to width ratio of a rectangular flake material.

assembly: A number of parts, materials, subassemblies, or combinations thereof placed together or joined using mechanical, adhesive or other methods.

assembly drawing: A document showing the physical relationship of two or more parts, materials or a combination of parts and subassemblies required to form an assembly of a more complicated nature.

assembly time: The interval between starting to bring parts, materials and other items physically together and the completion of the assembly process, such as the time from application of adhesive to cure of an entire assembly.

atactic: A random arrangement, such as the side chain locations off a polymer backbone.

attenuation: A transformation process for making an item thin; as the formation of fibers from molten glass. A reduction in amplitude.

autoclave: An enclosed vessel in which an inert atmosphere usually exists. Materials, both thermoplastic and thermoset are processed through the application of pressure, heat and vacuum assist. Chemical reactants may be vented from the mold through the autoclave wall.

autoclave bonding: The combination attachment and bonding of metallic, composite or both materials and/or parts using an autoclave to apply pressure, usually 50 to 200 psi (340 to 1380 KPa) at temperatures needed to cure adhesives. Bonding accomplished under disposable or reusable vacuum bag. Cured (honeycomb or foam) bonded structures usually require pressures of 50 psi or lower.

autoclave molding: A mold- or part-making process in which, after winding, wrapping, layup or placement of composite material, an entire assembly is placed in a heated autoclave. Applied pressure of 50 to 200 psi (340 to 1380 KPa) results in removal of reactants, volitiles and compacts, consolidates as well as cures the laminate of greater density. Bleeder, breather and other materials are used for this process.

automatic mold: A mold for injection or compression molding that repeatedly goes through the entire cycle, including ejection, without human assistance. Sometimes building stages (progressive), that perform functions completely with automatic sequence including hardware or fastener insertion to part ejection. Automated molding operations include resin transfer, compression, transfer, injection and others.

automatic press: A pneumatic or hydraulic press for compression, injection or other molding that can continually operate under electrical, mechanical, hydraulic, pneumatic or other control combinations.

autoxidation: The continuation self-sustained oxidation after initial exposure and termination of exposure to some oxidizing agent.

average molecular weight: A typical synthetic polymer is a mixture of various length molecular chains. The average molecular weight is determined by the polymer viscosity measured when a solution containing all the polymer chains is at a specific temperature, i.e., high viscosity/high molecular weight.

axial load: Load acting in compression or tension along a structural straight members long axis.

axial strain: A linear strain along or in a plane parallel to the longitudinal axis of a specimen.

axial stress: A stress of compression or tension resulting from applying a lengthwise axial load to a structural straight member.

axial winding: In filament winding reinforced composites, the filaments are parallel to the rotation axis with a 0° helix angle.

axis, symmetry: Being symmetric about an axis passing through the center of gravity of a material with all the surrounding area elements isotropic in the plane normal to the axis.

azeotropic mixture (axeotrope): A liquid or vapor mixture of two or more substances that behave like a single substance, i.e., liquid or vapor have the same composition.

B

back draft: An area of taper or depression in or on an otherwise smooth-drafted model or mold surface that causes a molded or shaped copy to interfere with the surface during removal; an obstruction in the taper which would interfere with the withdrawal of the model or mold.

back plate: Support plate in injection molding used to support the cavity blocks, bushings, guide pins, etc.

back pressure: The resistance through viscosity of a material to continued flow when a mold is closing. During extrusion it is the resistance to the forward flow of molten material, thereby resulting in improved melt mixing.

back taper: *See* back draft.

backing: In a prepreg, the film upon which the impregnated reinforcement lies. In a coated abrasive, the support material (cloth, paper, etc.) upon which the coating lies.

baffle: In an air circulating oven, a surface used to direct air flow. Also, a device used to restrict or divert fluid through a channel or pipe line.

bagging: *See* bag molding.

bag molding: A technique of molding reinforced laminate and other plastic composites using a flexible cover (disposable or reusable bag) over a rigid mold. The composite material is placed in or on the mold and covered with the bag which is usually sealed along the edge to the mold. Pressure is then applied by vacuum, autoclave, press, or by inflating the bag, thereby compressing and compacting the material to be molded against the mold surface.

bag side: The side of the part that is against the vacuum bag.

bag side tool: A mold or tool placed on top of part of the material being molded, but just under the bag.

bakelite: Once a trade name of the Bakelite Corporation used to denote phenolic resin systems.

baking: Process of curing or drying a material at a temperature in excess of 65°C resulting in ultimate properties.

balanced composite: A laminate composite layup that is symmetrical with relation to the midplane of the laminate.

balanced construction: In plywood or composites which have an odd number of plies, there is symmetrical balance of plies on both sides of the centerline.

balanced design: The design of a winding pattern in filament-wound reinforced composites resulting so that all stresses in all filaments are equal.

balanced-in-plane contour: For filament wound parts, a head contour where the filaments are oriented within a plane and the radii of curvature adjusted to balance the stresses along the filaments with the pressure loading.

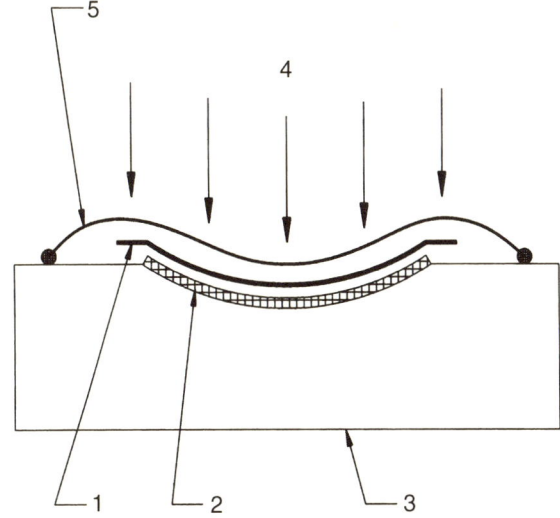

1–semirigid bag side tool
2–part material
3–female tool
4–atmospheric and/or autoclave pressure
5–vacuum bag

Bag side tool

balanced laminate: A composite laminate that is symmetrical about the centerline. All laminate at angles other than 0° and 90° occur only in ± pairs (not necessarily adjacent).

balanced runner: For injection molding, a system designed to place all cavities at the same distance from the sprue.

balanced twist: The twist arrangement in a plied yarn or cord which will not cause twisting on itself when the yarn or cord is held in the form of an open loop.

ball mill: Small machine usually with rubber-faced rollers that is used to rotate conical or other containers for the purpose of mixing, grinding, or dissolving solid resins in organic solvents. Ceramic balls are added to the ingredients inside the container to provide friction and movement.

banbury mixer: Used to compound materials. Composed of a pair of contra-rotating rotors which masticate the materials to form a homogeneous blend.

band density: For filament winding, the quantity of reinforcement per inch of bandwidth, expressed as strands (or filaments) per inch.

band width: For filament winding, the width of the reinforcement as it is applied to the mandrel surface.

barcol hardness: A value obtained by measuring the resistance to penetration of a sharp steel point under a spring load. The hardness value is often used as a measure of the degree of cure of a plastic.

barcol impressor: Barcol hardness measuring instrument.

bare glass: Glass (yarns, rovings, fabrics) without sizing or finish.

barrier coat: An exterior coating applied over a surface. Provides protection.

basal plane: A plane perpendicular to the principal "c" axis in a tetragonal or hexagonal structure.

base: The glass fiber, paper or other reinforcing material impregnated with a resin system in the forming of laminates; an insulating support for an electrical printed pattern.

batch: An industrial quantity of material formed during the same process or in one continuous process and having identical characteristics throughout. Also called a lot.

batt: Felted fabrics or structures built by the nonweaving or knitting interlocking action of fibers themselves.

B-basis: The "B" mechanical property value is the one above which at least 90% of all of the values are expected to fall, with a confidence level of 95%.

beam: A spool, for use in weaving or similar processing operations, on which is wound a number of parallel ends of single or plied yarns. Also a structural member or support.

bearing area: Diameter of the hole times the thickness of the material.

bearing strain: The stretch or deformation strain for a sample under bearing load.

bearing strength: The maximum bearing stress to be sustained. The bearing stress at that point on the stress-strain curve where the tangent is equal to the bearing stress divided by n percent of the bearing hole diameter.

bearing stress: An applied load in pounds divided by the bearing area.

bending moment: Stress that changes the curvature of a plate or beam.

beta gage: A thickness measuring device or gage consisting of two facing elements, an emitting source and detector. Sheet thickness is a measure of beta rays being absorbed when material is passed between the elements.

bias cut: To cut material 45° from the weave pattern.

bias fabric: Warp and fill fibers in a fabric are at an angle to the length.

biaxial: Where both normal components of stress or strain are acting.

biaxial load: A loading condition of a pressure vessel under internal pressure and with unrestrained ends. Also, loading condition in which a laminate is stressed in at least two different directions in the plane of the laminate.

biaxial winding: A type of filament winding in which the helical band is laid in sequence, side by side, with the crossover of fibers eliminated.

bidirectional laminate: A reinforced plastic cross-laminate with the fibers oriented in various directions in the plane of the laminate. *See also* unidirectional laminate.

billet: A small hot worked or cast ingot of material.

binary alloy: Alloy containing two component elements.

binder: Resin or cementing constituent or a plastic compound which holds the other components together; an agent applied to glass mat or performs to bond the fibers prior to laminating or molding.

bismaleimide (BMI): Polyimide type resin system that cures by an addition reaction, avoiding formation of volatiles, and has temperature capabilities between those of epoxy and polyimide. (200°C to 250°C)

bisphenol A: A condensation product formed by reaction of two (bis) molecules of phenol with acetone. It is a standard polyhydric phenol resin intermediate along with epichlorohydrin in the production of epoxy resins.

bladder: For filament winding structures it is the elastomeric lining for the containment of hydroproof or hydroburst pressurization medium.

blanket: A flexible bag process where fiber or fabric plies have been laid up in a complete assembly and placed on or in the mold all at one time. Also, the form of bag in which the edges are sealed against the mold.

blanking: The cutting (die cutting) of flat sheet material to shape using a sharp edged punch while it is supported on a mating die or flat surface.

bleed: An undesired movement of certain materials or color in adhesives (sometimes plasticizers) to the surface of the bonded article (especially plastics) or into adjacent material. Also, to evacuate resin, air or gases from a laminate molding assembly during the curing cycle.

bleeder: Bleeder fabrics designed to absorb excess resin and reactants during compaction.

bleeder cloth: A nonstructural, disposable tooling-aid layer of woven or nonwoven material, not a part of the composite layup, that allows excess gas reactants and resin to escape during cure. Removed from the part after curing.

blending: Thorough dry or wet mixing of various ingredients of a molding composition, or of several batches of the same type, to ensure uniform distribution of all particles.

blister: Undesirable dome and rounded elevation defect of the surface of a plastic, whose boundaries may be more or less sharply defined, and usually caused by entrapped gas or air beneath the surface.

block copolymer: A linear copolymer in which the backbone consists of regions or blocks of one monomer along with regions or blocks of another monomer.

blocked curing agent: A curing agent or hardener rendered unreactive. It can be reactivated by physical or chemical means.

blocking: Undesired sticking or adhesion between touching layers or particles of a material, that occurs under moderate pressure during storage or use.

bloom: A visible local exudation or finish change on the surface of a plastic or mold. Bloom can be caused in a plastic by a lubricant or plasticizer or by atmospheric contamination on a mold.

blow molding: A plastic shaping method in which a hollow tube parison is forced into the shape of the mold cavity by internal air pressure.

blow rate: Speed at which air enters the parison during the blow molding cycle.

blowing: Rising and expanding of a material during a foaming operation.

blowing agent: Expanding additive for resins to be foamed. When heated to a specific temperature, it decomposes to yield a large volume of gas that creates foamed plastic cells.

blown sand: In sand casting, a sand and resin-binder mixture blown, or packed in the tool to create a mold of the desired part.

blueing: A blue oxide film type mold blemish which occurs on the polished surface of a mold as a result of the use of abnormally high mold temperatures.

blush: Formation of whitish surface on a coating applied during a period of high humidity causing condensation on the surface by the fast evaporation of the coating solvent.

body putty: A paste like polyester or epoxy mixture of plastic resin and talc used in repair of metal auto and other surfaces.

bond: The union or uniting of materials by adhesives, sharing of electrons or other means.

bondline: Two or more materials or items are joined along this line.

bonded joint: Location where two or more adherends are bonded with a layer of adhesive. A lap joint results in overlaps, a scarf joint, with a matched tapered section and a stepped joint, through steps.

bonding: To attach materials with adhesive compounds.

bonding angle: The angle or connection where several plies of reinforcement and resin used to connect two parts of a laminate lies, usually at right angles to each other.

bond ply: A ply of prepreg material that is placed against the honeycomb core.

bond strength: The unit load or stress applied in tension, compression, flexure, peel, impact, cleavage, or shear, required to separate and break an adhesive assembly with failure occurring in or near the plane of the bond. The amount of adhesion or adherence.

boron fiber: A chemical vapor deposited fiber usually of a low density, high tensile strength, tungsten-filament core with elemental boron vapor deposited on it.

borsic: Silicon carbide coated boron fibers.

boss: Projection on a plastic part designed to add strength, facilitate alignment during assembly, to provide for fastenings, and permit attachment to another surface or part.

bottom plate: A steel plate fixed to the lower section of a mold. It is often used to join the lower section of the mold to the platen of the press.

bow: A condition of longitudinal curvature in molded parts or material.

braiding: Weaving fibers into a tubular shape about an axis rather than a flat fabric.

branched: A chemistry term that refers to side chains attached to the main original chain at irregular intervals.

branched polymer: In the molecular structure of polymers, a main chain with attached side chains, as opposed to a linear polymer.

breakdown voltage: The voltage required to cause the failure of a composite insulating material when under specific testing or operating conditions.

breaker plate: Perforated plate located at the rear end of an extruder head that often supports the screens to prevent foreign particles from entering the die.

breaking extension: Elongation necessary to cause rupture of the test specimen. The tensile strain at the moment of rupture.

breaking factor: Breaking load divided by the original width of the test specimen, expressed in pounds per inch.

breaking length: Measure of the breaking strength of yarn in inches. The length of a specimen whose weight is equal to the breaking load as a function of density.

breakout: Fiber separation or break on surface plies at drilled, machined, etc., edges.

breather: A loosely woven or nonwoven porous material, such as a fabric or mat, placed under the vacuum bag to facilitate removal of air, reactants, moisture, and volatiles by providing a path to the vacuum port, during the cure cycle.

breathing: Permeability to air of plastic sheeting or the opening and closing a mold to allow gases to escape.

bridging: A condition where one or more plies of impregnated fabric or prepreg span a radius, step, or chamfered edge of core without full contact.

brittle fracture: Fracture failure resulting with little or no plastic deformation.

brittleness: Quality of a material leading to crack propagation without appreciable plastic deformation.

broad goods: Fibers woven to form fabric wider than 12 inches and up to 1270 mm (50 in.) wide. It may or may not be impregnated with resin and is usually furnished in rolls of 25 to 140 kg (50 to 300 lb).

brushability: The adaptability of a material to application by brushing.

B-stage: Partial, intermediate or precured state of a thermosetting resin, allowing for easier storage, handling and processing. It is the stage where the system has been permitted to start the curing reaction without allowing the reaction to proceed to completion. B-staging increases resin viscosity, and minimizes resin flow during press or compression molding operations.

bubble: A spherical or internal void of air or other gas trapped within a plastic.

bubbler: In injection molding, a method of uniform cooling where a stream of cooling liquid flows continuously into a cooling cavity equipped with a coolant outlet that is normally positioned at the end opposite the inlet.

buckling: Composite buckling is a failure mode characterized generally by an unstable lateral deflection due to compressive or shear action on the structural element involved.

bulk bensity: Density of a loose granular, nodular molding material expressed as a ratio of weight to volume.

bulk factor: Reciprocal of fiber volume. Used to calculate lamina thickness. Ratio of the volume of a molding compound of powdered composite to the volume of the solid molded piece.

bulk modulus: The ratio of the hydrostatic pressure P to the volume strain in pounds per square inch.

bulk molding compound (BMC): Thermosetting resin system mixed with strand reinforcement, fillers, etc., into a viscous molding compound for compression or injection processing. *See also* premix and sheet molding compound.

bundle: Term for a collection of essentially parallel filaments or fibers.

bundle strength: Strength obtained from a test of parallel filaments with or without organic matrix and often used in place of the tedious monofilament test.

burn: To undergo combustion or a discoloration, distortion, or destruction of a molder or pultruded surface as a result of thermal decomposition.

burned: Showing evidence of thermal decomposition through some charring, discoloration, distortion, or destruction of the surface of the plastic.

burning rate: Term describing the tendency of plastic articles to burn at a given rate at given temperatures. Certain plastics, burn readily at low temperatures, and others will melt or disintegrate without actually burning or will burn only if exposed to direct flame and are referred to as self-extinguishing.

burst pressure: The pressure corresponding to the final failure of a pressure vessel.

burst strength (bursting strength): Measure of the ability of a material to withstand internal hydrostatic or gas dynamic pressure without rupture. Hydraulic pressure required to burst a vessel of given thickness.

bushing: The outer rings of any type of a circular tubing or pipe die which forms the outer surface of the extruded tube or pipe. Special extra heavy load-

butt fusion

carrying short cylinder inserted in bolt or pinholes. An electrically heated alloy container encased in insulating material, used for melting and feeding glass in the forming of individual fibers or filaments.

butt fusion: Joining method for pipe, sheet or other similar forms of a thermoplastic resin wherein the ends of the two pieces to be joined are heated to the molten state and then rapidly pressed together to form a homogeneous bond.

butt joint: A joint in which parts are joined end to end with no overlap.

butt wrap: Tape wrapped around an object in an edge-to-edge condition.

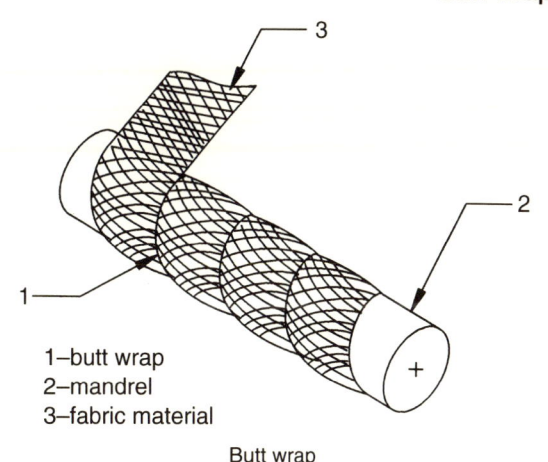

1–butt wrap
2–mandrel
3–fabric material

Butt wrap

C

calender: A machine and process used to produce a smooth finish and a desired dimensional thickness for sheet material by passing it between sets of pressure rollers.

caprolactam: A cyclic amide containing six carbon atoms.

carbon: The element that provides the backbone for all organic polymers. Graphite is a more ordered form of carbon; diamond is the densest crystalline form of carbon. Most of the high strength "graphite" fibers are actually carbon fibers.

carbon black: A black pigment or filler produced by the incomplete burning of natural gas, oil or other hydrocarbons.

carbon/carbon: A solid or foamed composite material consisting of carbon or graphic fibers in a carbon or graphic matrix.

carbon fiber: Light weight fibrous materials that are stiff, high strength and produced by the pyrolysis of organic precursor fibers, such as rayon, polyacrylonitrile (PAN), and pitch, in an inert environment.

carbonization: The pyrolyzation process performed in an inert atmosphere at temperatures from 800–1600°C.

cartridge heater: An electrical heater used to heat injection, compression, transfer molds, and associated hardware.

case harden: Harden the surface of a piece of steel.

casein: A protein material precipitated from skimmed milk.

cast: The process that describes forming an object by pouring a liquid polymer into a mold where it hardens to a finished shape.

casting: Process used to form film, sheet or finished poured molded parts.

catalyst: A substance which changes the rate of a chemical reaction without itself undergoing permanent change in its composition; it is referred to as an initiator if it becomes a part of the chain reaction. It speeds up the cure of a compound when added in small quantity as compared to the amounts of primary reactants. *See also* promotor, hardener, inhibitor, and curing agent.

catastrophic failure: Unpredictable mass failure of a physical, mechanical, thermal, electrical or other nature.

catenary: In fibrous or roving materials, the tendency of some strands in a taut horizontal direction to sag lower than the others.

cationic polymerization: The polymerization of any compound or mixture.

caul: A sheet of material (metal, wood, plastic, elastomer, other), used as a tooling aid for equalizing pressure during the laminating process.

caul pad: Similar to caul plate except usually elastomeric, free of surface defects, the same size and shape as a composite layup (and under the vacuum bag) used immediately in contact with the layup during the curing process to transmit normal pressure and temperature, and to provide a smooth surface on the finished laminate.

caul plate: A smooth bag-side or press platen plate used in contact with the layup during curing to transmit normal pressure and to provide a smooth surface to the finished laminate.

cavity: Space or spaces (single or multiple cavity) inside a mold in which a resin system, molding compound or other material is injected or poured. The female portion of a mold, that portion of the mold which encloses the molded article.

cell: In a honeycomb core material, a cell is a single (usually hexagonal) opening or part. The cell size is the diameter of a circle within a cell of honeycomb core.

cellular plastics: Materials with connected or unconnected cell structure throughout.

cellulosics: A broad family of hard materials that include cellulose acetate, butyrate, propionate, and ethyl cellulose. They possess broad property ranges.

cementitious: Compounds that act like cement and concrete products.

centerless grinding: A machining technique for making circular cross-section parts. The part is fed without mounting. The material is ground between wheels, (external or internal) which rotate at different speeds, the faster wheel being the abrasive wheel.

centipoise: Unit of viscosity conveniently and approximately defined as and compared to the viscosity of water, at room temperature.

Comparison Table

Liquid	Viscosity in Centipoises
Water	1
Glycerin	1,000
Molasses	100,000

centrifugal casting: A manufacturing method forming thermoset or thermoplastic resin materials into pipe or cylinders. The material resin is placed in a rotatable container, heated (or not heated) by the transfer of heat through the walls of the container and rotated so that the centrifugal force will force the molten resin to conform to the configuration of the interior surface of the container.

ceramic(s): Inorganic, nonmetallic materials, reinforced or non-reinforced, high compressive strength and temperature resistant. Low flexural and tensile strength. Various grades available for both room temperature and high temperature processing. Can be used to form molds, tools and parts.

cermet: Composite powder metallurgy material consisting of two components (inorganic-ceramic compound, and metallic binder).

C-glass: Chemical grade glass fibers (65% silica, 14% calcium oxide, 8% sodium oxide) with good resistance to chemicals and corrosive environments.

chain length: Indicate degrees of polymerization by the length of the stretched linear macromolecule.

chain scission: The breaking or severing degradation and loss of a side group or shortening of the overall chain.

chalking: Dry, chalk-like appearance of or deposit on the surface of a plastic material or paint surface usually caused by degradation.

charge: The amount (measurement or weight) of materials, (liquid, preform, or powder), used to load a mold during a molding cycle at one time.

Charpy impact test: A pendulum-type destructing impact resistance test for shock loading in which a centrally notched sample bar is held at both ends and broken by striking with a single blow, the back face in the same plane as the notch.

charring: A resulting condition after heating a composite in air reducing the polymer matrix to ash.

chase: The main body of the mold which contains single or multiple cavities, cores, pins, bushings, etc.

chemical milling: A machining method using chemicals for removing metal.

chemical polishing: A chemical treatment process to improve a metal surface.

chemical vapor deposition (CVD): A single or multi-coating process in which desired reinforcement material is deposited from a vapor onto a single item or continuous surface such as fiber.

chill: To cool a mold tool or part by water, air or other means.

chopped strand mat (CSM): Usually glass fiber reinforcement consisting of strands of fibers chopped into short lengths and bonded to form a planar mat.

CIL (flow test): A rheology determining method measuring flow properties of thermoplastic resins developed by Canadian Industries Limited.

circuit: In filament winding this is one complete traverse of a winding band through the machine fiber feed mechanism.

circuit-board: A sheet of insulating material which has a printed wiring (circuit) on one or both surfaces. The printed wiring pattern may be formed by adding wire, adding or removing metallic foil or other means, including through hole and board conductors.

circumferential ("circ") winding: In filament winding a "level winding" 90° with the filaments essentially perpendicular to the axis.

clad metal: A composite of metal containing two or three layers, bonded, plated, rolled or connected by other means.

clamping plate: A mold plate fitted to a mold or tool and used to fasten the mold to the machine.

clamping pressure: The pressure applied to an injection or transfer mold to hold it closed.

class A surface: High-gloss finish specified for the automotive industry.

clearance: Controlled distance, gap or space between two adjacent surfaces.

closed cell foam: Cellular plastic material without interconnecting passages or cells.

closure: Complete coverage of the mandrel in filament winding.

 Plain
 Crow Foot
 Basket
 Satin Weave

Cloth patterns

cloth: A woven fabric product made from continuous yarns or tows of fiber.

coalescence: The fusing of an aqueous liquid film upon evaporation of the water.

cocure: Simultaneous bonding and curing, low cost method of manufacturing a composite laminate with another material or parts (honeycomb core, laminate and stiffeners).

coefficient of elasticity: Reciprocal of Young's modulus in a tension test. *See also* modulus of elasticity.

coefficient of expansion: Fractional change in dimension of a material for a unit change in temperature. *See also* coefficient of thermal expansion.

coefficient of friction: Measure of the resistance to sliding or rolling and moving one surface in contact with another.

coefficient of linear expansion: *See* coefficient of expansion.

coefficient of thermal expansion (α): Change in volume per unit volume produced by a one degree rise in temperature.

cohesion: The ability of a single substance to adhere to itself by the internal attraction of molecular particles toward each other and thereby resist separation in any case.

coke: A carbonaceous residue resulting from pyrolysis of pitch.

cold flow: The physical dimensional change and "creep" distortion in materials under a continuous load and at temperatures within the working range.

cold molding: A room temperature composite molding method.

cold pressing: A pressure bonding operation taking place without the application of heat.

cold-setting adhesive: A room temperature setting resin system adhesive.

collet: A metal band; the drive wheel that pulls glass fibers from a bushing.

collimated: Parallel.

collimated roving: Special parallel roving that has been made using a different process from standard roving.

colloidal: State of suspension in a liquid where the suspended particles are dispersed but not dissolved.

colorant: Plastics pigment or dye.

combined stresses: All active stress components.

commingled: The close intermixing of (usually) reinforcement and thermoplastic or other matrix fibers within a single tow.

compaction: A densification process for removing entrapped air and volatiles during the layup process by the application of a temporary vacuum bag and vacuum.

compatibility: When combined together, the ability of two or more substances to form a homogeneous composition.

complex dielectric constant: Vectorial sum of the dielectric constant and the loss factor.

complex shear modulus: Vectorial sum of the shear modulus and the loss modulus.

complex Young's modulus: Vectorial sum of Young's modulus and the loss modulus.

compliance: Tensile compliance; the reciprocal of Young's modulus; shear compliance; the reciprocal of shear modulus.

composite: A homogeneous material of two or more materials of different physical and mechanical properties combined to form shapes or structures with improved specific combined properties but retain their original identity.

compound: A mixture of a resin system with other ingredients, such as fillers, reinforcement, etc.

compounding: The mixing process of combining all the materials necessary for making the complete compound.

compression mold: A mold which is open when the material is introduced and which shapes the material with or without heat and by the pressure of closing.

compression molding: A thermoset material molding process where the molding material is placed in the open mold, the mold is closed, heat and pressure applied until the material is cured.

compression set: The percent a material deforms from the original shape after an applied load is removed.

compressive modulus: Compressive stress to compressive strain ratio.

compressive strength: Ability of a material to resist a crushing force. The pressure load is measured at failure.

15

concentricity: The dimensional relationship of all inside dimensions of a diameter.

condensation: Chemical reaction in which two or more gas or vapor molecules combine with the result being reduction to a liquid or solid.

condensation agent: Chemical compound which, becomes part of the polycondensation reaction and acts as a catalytic.

condensation polymer: Polymer formed by polycondensation.

condensation polymerization: A chemical reaction in which two or more molecules combine, with the separation of water or some other simple substance. If a polymer is formed, the process is called polycondensation. *See also* polymerization.

condensation resin: Resin formed by polycondensation.

conditioning: Subjecting materials, parts, and products to specific environments prior to test or use.

conductive materials: Copper, nickel, silver, graphite, and other metallic-based compounds used as coating on nonconductive substrates for EMI/RFI shielding or conducting current and/or heat. Can be cast or poured as encapsulants.

conductivity: Reciprocal of volume resistivity. The electrical or thermal conductance per unit volume (cube) of any material.

consistency: Used to describe how a material will flow (viscosity) when shearing stresses are applied to it.

consolidation: In metal matrix or thermoplastic composites, a processing step in which fiber and matrix are compressed by one or several methods to reduce voids and achieve a desired density.

constituent materials: Materials that make up a composite material; e.g., polyester/fiberglass are the constituent materials of a polyester/fiberglass laminate.

contact adhesive: An adhesive when applied to both adherends and allowed to become partially dry, will bond when the adherends are brought together with little sustained pressure, but will not bond to other materials.

contact molding: A molding process where a resin system is used to impregnate a reinforcement placed on the surface of an open mold. Materials cure at room temperature or by heat.

contact pressure resins: Liquid resins which when used for laminating and bonding thicken or polymerize on heating.

contaminant: An undesirable foreign substance.

continuous filament: An individual rod of glass of small diameter, which is flexible and of great or indefinite length.

continuous filament yarn: A single, small diameter indefinite length fiber (rod of glass, carbon, aramid, other).

cooling fixture: A tooling aid used to maintain the shape and dimensional accuracy of a cast or molded material formed into a part (bonded assembly, laminate, etc.) after it is removed from the mold or tool.

copolymer: A chemical linked long-chain polymeric system which contains two or more monomeric units.

copolymerization: Copolymers (two or more monomers) simultaneously polymerized.

core: The center (honeycomb or other) material to which metal, composite or other faces are attached and bonded. Also cooling or heating a channel in a mold.

core-crush: Collapse and compression of the core at or along an edge or internal area (local identation).

cored mold: A mold with channels or incorporating passages for heating or cooling.

core separation: A disbond, partial or complete breaking of the faces or skins from the core.

core splice: Joint or connection of one or more core pieces.

corrosion: Chemical or physical degradation of metallic or nonmetallic surfaces of a material.

count: For fabric, number of warp and filling yarns per square inch, for yarn, size based on relation of length and weight.

coupling agent: Any chemical substance designed to be added to a matrix resin system, added to and react with both the reinforcement and matrix, to promote a stronger bond at the interface.

coupon: Another name for a test specimen (tensile, flexural, etc.).

cowoven fabric: A reinforcement fabric woven with two different types of fibers.

crack: A fracture or failure of a material (a separation).

crazing: Fine surface crack in a single or multiple form.

cream time: Length of time between mixing and pouring (beginning of foam phase) of a mixed thermoset foam system.

creel: An attachment of a braiding, filament winding or weaving machine to hold roving balls (spools) in order to unwind or feed fiber properly.

creep: The change (cold flow at room temperature) in dimension of a plastic under load over a period of time, not including the initial instantaneous elastic deformation.

creep, rate of: The slope of the creep-time curve at a given time. Deflection with time under a given static load.

crimp: Waviness of a fiber or fabric per unit length.

critical buckling loads: Lowest loads at which buckling occurs.

critical strain: The strain at the yield point.

cross-laminate: A laminate where layers of material are oriented at right angles to the remaining layers.

cross-linking: In a polymer molecule, setting up of chemical links (primary valence bonds) between the molecular chains.

cross ply: Not uniaxial; having plies oriented in different directions.

cross-ply laminate: A 0°/90° laminate ply.

crosswise direction: Crosswise is an arbitrary direction at right angles to the lengthwise direction of any material equal in strength in both directions. Crosswise also refers to cutting a specimen and to applying load. Crosswise is the weaker direction in a material or unequal directional strength.

crowfoot weave: A warp-faced weave in which the binding places are arranged with a view to produce a smooth cloth surface free from twill, with a regular interlacing shift (satin weave).

crystal: Solid composed of atoms, ions, or molecules arranged in a pattern that is repetitive in three dimensions.

crystalline plastic: A polymeric material with an internal structure where the atoms are arranged in an orderly three-dimensional configuration.

crystallinity: Atoms arranged in an orderly, three-dimensional pattern in a molecular structure.

crystallite: Small crystal. (micro- or sub-microscopic)

C-scan: Nondestructive ultrasonic testing technique for finding flaws in composite.

C-stage: Final stage in the reaction of a thermosetting resin system. The resin system in a fully cured thermoset molding is at this stage.

cull: Transfer molding machine material remaining in a transfer chamber after the mold has been filled. Also reject materials or products.

cure: To permanently change the properties of a thermosetting resin system by chemical reaction, (condensation, ring closure, or addition). Accomplished with or without heat or pressure using hardeners or curing (cross-linking) agents.

cure cycle: Time/temperature/pressure profile used when curing a thermosetting resin system or prepreg.

cure monitoring, electrical (dielectric): Measuring the resin system cure profile using electrical techniques to detect changes in material electrical properties.

cure stress: An internal composite structural laminate stress caused by matrix resin system shrinkage, mold and tool material differences during the cure cycle. Also caused by different thermal coefficients of expansion of the materials in the laminate.

cure temperature: Temperature at which a thermoset resin system reaches final cure.

curing agent: Cross-linking agent, hardener, catalytic or reactive agent added to a resin to cause polymerization.

curing time: Total period of time from the start of the molding cycle (heat or pressure—both) until a part is removed from the mold or tool until fully cured (total cross-linking).

cycle: Full time sequence in one molding operation.

D

dam: The transition or boundary along the edge of a mold or a tool used to prevent the resin system from flowing out of a laminate or the vacuum bag from wrinkling along the edge.

damage tolerance: The measure of a structure's ability to resist crack growth rate after exposure or impact due to sudden load such as a projectile.

damping: The dissipation of energy during deformation of a material.

dash-pot: A shock absorber for damping down vibration. Formed from a pot of air or fluid with a piston connected to the part to be damped.

database: A comprehensive collection of information.

daylight: The clearance or open distance between two platens of a molding or bonding press.

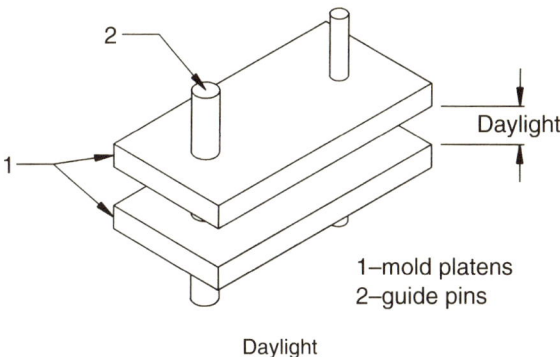

Daylight

deaerate: Remove air by vacuum or other means.

debond: An unplanned or deliberate area of separation within or between laminate plies or a bonded joint.

debulk: A step in the laminating process where laminate thickness is reduced by application of pressure. Achieved by removing volatiles, air, vapor, and other reactants.

deep-draw: A molding process where a mold core is long in relation to the molded part wall thickness.

deflashing: A finishing process where excess material resulting from the molding cycle (flash) is removed by grinding, filing, sanding, tumbling, etc.

deflection: Displacement or movement of a structure, beam, etc., from it's original shape or position.

deflection temperature under load: The "heat distortion temperature" at which a beam has deflected a given amount under load.

deformation: Any changes in size, form or shape of a body as a result of externally applied stresses, temperature change, and moisture absorption.

deformation under load: The "cold flow," "creep" or dimensional change of a material under a load, stress or forces for a specific period of time.

degassing: Removal of gases, volatiles and air using vacuum or other means.

degradation: A loss of physical or other properties due to excessive heat, light exposure, aging and corrosion.

degree of polymerization (cross-linking): The measure of the cross-linking of a polymer.

delaminate, delamination: Separation, debonding, of a laminated plastic material along the plane of its layers or laminate plies.

deliquescent: The ability of a material to attract moisture from the surrounding environment.

denier: The unit weight for a yarn and filament numbering system in which the yarn number is equal numerically to the weight in grams of 9000 meters.

densification: The consolidation or compaction of a loose or bulky material (debulking).

densitometer: An instrument that is used to measure the amount of light transmittance and absorption a material experiences.

density: Mass or weight per unit volume of a substance expressed in grams per cubic centimeter, pounds per cubic foot, etc.

deposition: An additive process for applying a material to a substrate by means of chemical, vacuum, electrical, screening, or vapor methods.

19

desiccant: Chemical substance used for drying and absorbing water vapor.

design allowables (by a test program): Material property allowable strengths, statistically defined.

design automation: Use of computer systems, programs, and procedures in the design process. The computer is responsible for the design process, the decision-making activity and data manipulation functions.

desizing: Process of removing sizing, (starch, lubricant) from gray (also greige) goods before applying special finishes or bleaches.

desorption: Process for removal of absorbed material from another material.

destaticization: A treatment process of plastics materials which will minimize their accumulation of static electricity.

deterioration: A reduction or negative change in the physical properties of a plastic material.

devitrification: Formation of crystals (seeds) in a glass melt occurring when the melt is too cold.

D-glass: High boron content glass for laminates requiring a precisely controlled dielectric constant.

die: *See* cavity.

die blades: Part of equipment used to adjust and to produce uniform thickness across film or sheet being produced.

die block: The part of an extrusion die which holds the forming bushing and core.

die body: That part of an extrusion die used to separate and form material.

die cutting: Clicking, blanking or cutting shapes from sheet stock with a shaped knife edge tool known as a "steel-rule die."

dielectric: A material with very little electrical conductivity. In radio frequency (RF) heating and curing the term dielectric is the material being heated.

dielectric constant: Ratio of the capacitance of an assembly of two electrodes separated by dielectric material to its capacitance when the electrodes are separated by air.

dielectric curing: Curing a thermosetting resin by the passage of an electric charge through the resin system.

dielectric heating: Heating of materials by placing a plastic to be heated (dielectric) in a high-frequency electric field.

dielectric insulating material: In radio frequency preheating, dielectric may refer specifically to the material which is being heated.

dielectric loss: Loss of energy of a dielectric placed in an alternative electric field.

dielectric loss angle: Dielectric phase difference between an angle of 90° and the dielectic phase angle.

dielectric monitoring: The process of monitoring or tracking the cure of thermosets by recording changes in their electrical properties during cure.

dielectric strength: Electric voltage gradient at which an insulating material fails, breaks down and is "arced through" (volts per mil of thickness).

differential scanning calorimetry (DSC): Process of measurement of the energy absorbed (endotherm) or produced (exotherm) as a resin system is cured as well as detecting loss of solvents and other volatiles.

differential thermal analysis (DTA): An analysis technique where a specimen and control sample are heated simultaneously. The difference in their temperature is monitored and provides information on relative heat capacities, residual solvents, changes in structure, etc.

diffusion bond: Process used to consolidate materials which require the application of heat and pressure.

dilatancy: Property characteristic of some materials by which the resistance to flow increases with agitation.

diluent: An ingredient usually added to another to reduce the viscosity and dilute the other.

dimensional stability: Shape retention in cast, molded, or formed parts.

dimer: Substance comprised of two molecules of a monomer.

dimorphism: The ability of a substance to exist in two different crystalline forms.

dip coating: Application of a plastic coating by dipping the item to be coated into a container of the appropriate resin system.

directional properties: Physical and mechanical properties that vary depending on the relation of the test axis to a preferred orientation.

disbond: A separation or adhesion failure with a bonded interface between two adherends.

dished: A warped or symmetrical distortion of a flat or curved section of a plastic object.

dispersion: Finely divided particles of a material in suspension in another substance.

displacement angle (filament winding): The advancement distance of the winding material on the equator after one complete circuit.

dissipation factor-electrical: Ratio of the power loss in a dielectric material to the total power transmitted through the dielectric.

distortion: In a solid, the change in shape, in fabric, the displacement of fill fiber from the 90° angle (right angle) relative to the warp fiber.

doctor blade or bar: A flat, straight piece of material used to meter and spread coating material to rollers

doctor roll of a coating machine, substrate, reinforcement surface, etc.

doctor roll: In adhesive or coating equipment, the roller device that rolls at a different surface speed, or in an opposite direction, and provides a wiping action for regulating the adhesive supplied to the spreader roll.

doily (filament winding): Planar reinforcement applied to a local area between windings especially where a cutout is to be made.

dome (filament winding): An integral end of a cylindrical container with extra wound reinforcement.

double-shot molding: Molding parts by using molded part number one as an insert for molded part number two.

doubler: Extra plies or layers of reinforcement, that provide stiffness and strength to fasten at that location.

double-back tape: Tape with adhesive and protective separator on both sides.

dry laminate: A laminate with an insufficient amount of resin system for complete wet-out and bonding of the reinforcement.

dry layup: An application process for prepreg (preimpregnated reinforcement) where layers are applied to male or female mold surfaces, usually vacuum bagged and cured (autoclave, other).

dry spot: A resin starved area.

dry strength: Strength of a composite laminate, specimen, adhesive joint or part immediately after specific conditioning and drying under special conditions.

dry tack: A property of certain adhesives to adhere to themselves on contact.

dry winding: Filament winding using preimpregnated roving.

drying time: Time period during which an adhesive on an adhered or an assembly is allowed to dry with or without the application of heat or pressure, or both.

ductility: Is the measure of plastic strain a material can withstand before fracture.

durometer: An instrument (usually hand held) used to measure an elastomer or plastic material hardness.

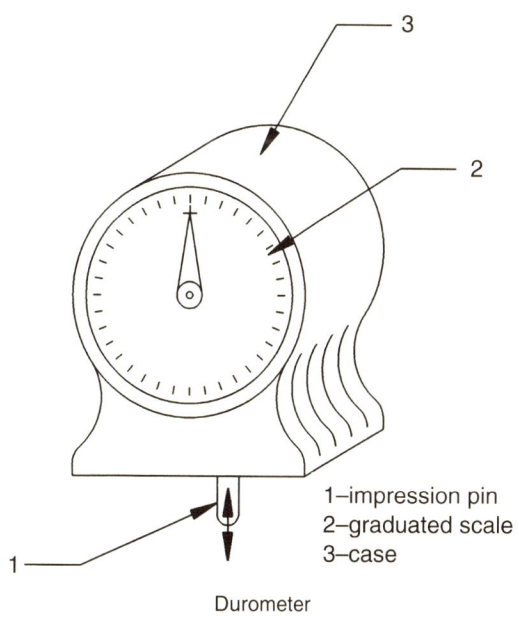

1–impression pin
2–graduated scale
3–case

Durometer

durometer hardness: Hardness of a material as measured by a durometer dial gage.

dwell: A pause in time usually associated with the application of pressure, or temperature, materials or equipment.

dye: Natural or synthetic organic chemical soluble in most common solvents, having good tinting or coloring strength.

dynamic modulus: The ratio of stress to strain under vibratory conditions in shear, compression, or elongation.

E

edge dam: Boundary support or ridge to prevent excessive run out from the laminate before or during cure.

edge injection: In resin transfer molding, the positioning of the inlet port such that resin flow is parallel to the laminate layers.

edge joint: A joint made by bonding the edge faces of two adherends.

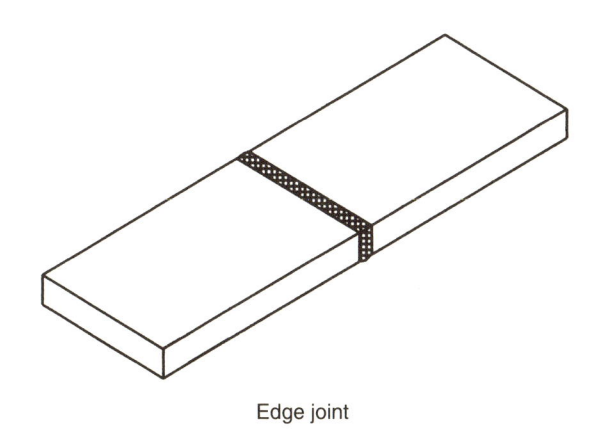

Edge joint

edge molding: The technique of molding the periphery of components to net finished size.

edgewise: In cutting, the application of forces in directions parallel to and actually in the plane of a sheet or specimen. For compression molded specimens of square cross section, the edge is the surface parallel to the direction of motion of the molding plunger. For injection-molded specimens of square cross section, this surface is selected arbitrarily. For laminates, the edge is the surface perpendicular to the laminate.

egg-box: The supporting substructure for a mold or tool, fabricated as an interlocking box assembly.

E-glass: Electrical insulation grade glass fibers (54% silica, 18% CaO, 14% alumina, 8% boria). The most common fiber for reinforced plastics.

ejection: The process of removing a molding from the mold impression; by mechanical means, by hand, or by the use of compressed air.

ejection plate: A metal plate used to operate ejector pins; designed to apply a uniform pressure to them in the process of ejection.

ejector pin: Pin or dowel used to eject molded articles from a mold.

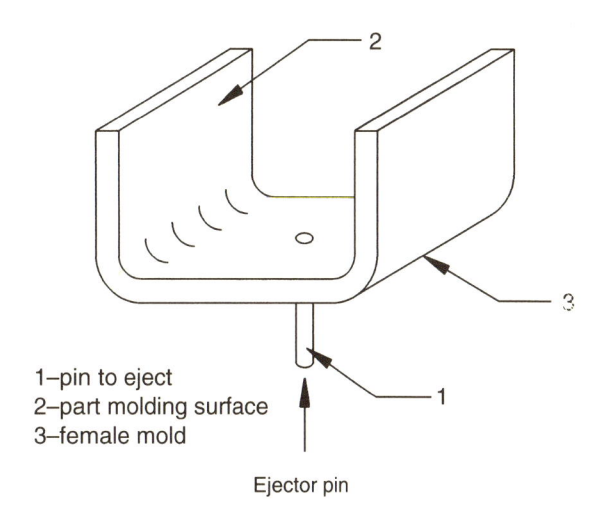

1–pin to eject
2–part molding surface
3–female mold

Ejector pin

ejector rod: Bar that actuates the ejector assembly when mold is opened.

elastic deformation: That part of the total strain in a stressed body which disappears upon removal of the stress.

elastic limit: The extent to which a material can be stretched or deformed before taking on a permanent set. Permanent set occurs when a material that has been stressed does not recover its original dimensions, as when a 12-in. piece of rubber that has been stretched becomes 13 in. long when relaxed.

elastic memory: The ability of a thermoplastic to return to its original shape when exposed to heat beyond its heat distortion point. A flat sheet that has been thermoformed to a new shape reverts to a flat sheet if sufficiently heated.

elastic relation: Fully reversible, single-valued stress-strain relation. Loading and unloading follow the same path; there is no hysteresis, or residual strain. Although nonlinear relation is admissible, the relation for composite materials is essentially linear.

elastic rotor: A form of screwless extrusion where melt is achieved mechanically between rotating members.

elasticity: The property of plastic materials by virtue of which they tend to recover their original size and shape after deformation.

elasticity, coefficient of: *See* coefficient of elasticity.

elastomer: A material which at room temperature stretches under low stress to at least twice its length and snaps back to the original length upon release of stress.

elastomeric: A rubbery material, either silicone or non-silicone bonded, that can be cast, fabricated or extruded to requirements.

elastomeric molding: A composite tooling/curing technique where a rubber compound (an elastomer) is placed in a confined space (between the tooling surfaces and the laminate) so that increasing the temperature causes the rubber to expand and apply pressure to the laminate being cured. Rubber materials have a coefficient of thermal expansion an order of magnitude greater than most composite tooling materials.

elastomeric tooling: A tooling system that uses the thermal expansion of rubber materials to form composite parts during cure.

electric discharge machining (EDM): A metal-working process applicable to mold construction in which controlled sparking is used to erode the work piece forming the finished molding surface.

electric surface resistance: The surface resistance between two electrodes in contact with a material is the ratio of the voltage applied to the electrodes to that portion of the current between them which flows through the surface layers.

electric surface resistivity: The ratio of the potential gradient parallel to the current along the surface of a material to the current per unit width of surface.

electrical dissipation factor: The ratio of the power loss in a dielectric material to the total power transmitted through the dielectric, or the imperfection of the dielectric. Equal to the tangent of the dielectric loss angle:

where f is frequency of applied voltage in hertz, C_p is equivalent parallel capacity, and R_p is equivalent parallel resistance.

electroform: Any mold or tool manufactured utilizing the electrode-position process.

electroformed mold: A mold made by electroplating metal on the reverse pattern on the cavity.

electromagnetic interference (EMI): An electromagnetic energy that causes interference in the operation of electronic equipment.

electroplating: Deposition of metals on certain plastics and molds for finish.

electrostatic discharge (ESD): A large electrical potential (4000 V or more) moving from one surface or substance to another. ESD is also an abbreviation for electrostatic dissipation.

elongation: Deformation caused by stretching; the fractional increase in length of a material stressed in tension. (When expressed as percentage of the original gage length it is called percentage elongation.)

elongation at break: Elongation recorded at the moment of rupture of a specimen, often expressed as a percentage of the original length.

embossing: Techniques used to create depressions of a specific pattern in plastic film and sheeting.

emissivity: The ratio of the total heat radiating power of a surface to that of a black body of the same area and of the same temperature.

emulsifier: A material which, when added to a mixture of dissimilar materials, such as oil and water, will produce a stable homogeneous emulsion.

emulsion: A suspension of extremely fine droplets of one liquid in another.

emulsion polymerization: The process of polymerization taking place in the presence of water to form a latex.

encapsulating: Enclosing an article in an envelope of plastic by immersing the object in a casting resin and allowing the resin to polymerize or, if hot, to cool.

end: The smallest commercially available bundle of glass fibers. A loosely bonded bundle or continuous fibres formed and sized together (of strand).

end count: An exact number of ends supplied on a ball or roving.

endothermic reaction: A reaction which is accompanied by the absorption of heat.

endurance limit: *See* fatigue limit.

engineering constants: Measured directly from uniaxial tensile and compressive, and pure shear tests applied to unidirectional as well as laminated composites. Typical constants are the effective Young's modulus, Poisson's ratio, and shear modulus. Each constant is accompanied by letter or numeric subscripts designating the direction associated with the property.

engineering plastics: Plastics that lend themselves to

envenomation: The process by which the surface of a plastic close to or in contact with another surface is deteriorated. Softening, discoloration, mottling, crazing or other effects may occur.

environment: The aggregate of all conditions (such as contamination, temperature, humidity, radiation, magnetic and electric fields, shock, and vibration) that externally influence the performance of an item.

environmental stress cracking: The susceptibility of the thermoplastic resin to crack or craze when in the presence of surface active agents or other environments. (ESC)

epichlorohydrin: The basic epoxidizing resin intermediate in the production of epoxy resins. It contains an epoxy group and is highly reactive with polyhydric phenols such as bisphenol A.

epoxide: A reactive group in which an oxygen atom is joined to each of two carbon atoms which are already united in some other way.

epoxide equivalent: The weight of a resin in grams which contains one gram equivalent of epoxy.

epoxy: Thermosetting resin made by polymerization of an epoxide.

epoxy molding compound: Compounds are mineral-filled powders which can be molded on compression or transfer molding presses.

epoxy plastics: Group of plastics composed of resins produced by reactions of epoxides or oxiranes with compounds such as amines, phenols, alcohols, carboxylic acids, acid anhydrides and unsaturated compounds.

epoxy resins: Thermosetting polymers cured by a ring opening mechanism. The epoxy (oxirane) group is a three-membered ring of two carbon atoms and one oxygen atom. The glycidyl group is a methylene group adjacent to an oxirane group.

equator: The line, in filament winding, described by the junction of the cylindrical portion and the end dome.

erosion: Destruction of metals or other materials by the abrasive action of moving fluids, usually accelerated by the presence of solid particles or matter in suspension. When corrosion occurs simultaneously, the term erosion-corrosion is often used.

ester: The reaction product of an alcohol and an acid.

ethyl alcohol: A colorless, volatile liquid; generally used for cleaning bond surfaces.

ethylene plastics: Group of plastics formed by polymerization of ethylene or by the copolymerization of ethylene with various unsaturated compounds.

eutectic: An alloy having the composition indicated by the eutectic point on an equilibrium diagram. An alloy structure of intermixed solid constituents.

exotherm: The liberation or evolution of heat during the curing of a plastic product.

exothermic: Indicating a reaction which liberates heat.

expandable plastics: Plastics that can be transformed to cellular structures by chemical, thermal or mechanical means.

expansion coefficient: Measurement of swelling or expansion of composite materials due to temperature change and moisture absorption.

extend: Add fillers of lower cost materials in an economy-producing endeavor; to add inert materials to improve void-filling characteristics and to reduce crazing.

extender: A material which, when added to an adhesive, reduces the amount of primary binder necessary.

extensibility: The ability of a material to extend or elongate upon application of sufficient force. Expressed as percent of the original length.

extensometer: A mechanical or optical device for measuring linear strain due to mechanical stress.

extraction: Transfer of materials from plastics to liquids with which they are in contact.

extrudate: The product or material delivered by an extruder, such as film, pipe, the coating on wire, etc.

extrusion: Process in which heated or unheated plastic compound is forced through an orifice, forming a continuous article.

extrusion coating: The coating of a resin on a substrate by extruding a thin film of molten resin and pressing it onto or into the substrate without adhesive.

F

fabric: A planar assembly of fibers interlaced by weaving, braiding or knitting. Nonwovens are sometimes included in this classification.
fabric, nonwoven: Material formed from fibers of yarns without interlacing.
fabric, woven: A planar assembly of fibers, filaments or interlaced yarns.
fabricated tooling: Any mold or tool manufactured by mechanically fastening, bonding or welding component parts.
fabric fill face: That side of the woven fabric where the greatest number of the yarns are perpendicular to the selvage.
fabric warp face: That side of the woven fabric where the greatest number of the yarns are parallel to the selvage.
fabric wrinkle: Condition where one or more plies of material are permanently formed into a ridge, depression, or fold.
face: The outer ply of a laminate.
fadeometer: Apparatus for accelerating fading and determining a material's resistance.
failure, adhesive: Rupture or separation of an adhesive at the adhesive adherend interface.
failure, cohesive: Rupture or separation of an adhesive bond within the adhesive.
failure criterion: Failure mode and description of composite materials subjected to complex states of maximum stresses or strains.
failure envelope: The limit of combined stress or strain state as defined by the failure criterion.
fair: To blend and join adjacent surfaces.
fairing: A structure or member, which is smooth and streamlines the flow of air or a fluid by reducing drag.
false twisting: An additive manufacturing process on yarn so that no net twist is inserted.
fan gate: A broad opening in the mold gate area between the mold runner and the mold cavity shaped like a fan.

fatigue: The reduction or complete failure of mechanical properties following repeated cycles of stress.
fatigue life: For a stated test, the number of cycles of stress or strain prior to material failure.
fatigue limit: Stress or point below which materials can be cyclically stressed any number of times without failure.
fatigue notch factor: The fatigue strength ratio of unnotched to notched specimens at an equal number of cycles.
fatigue notch sensitivity: Estimating the effect, fatigue strength or material life after putting a notch in a specimen and testing.
fatigue strength: The residual strength after being subjected to cyclic stress and fatigue.
fatigue stress ratio: Ratio of the minimum to maximum fatigue stress.
faying surface: Surfaces of materials in contact with each other and joined or about to be joined.
Feigenbaum, Armand: Quality guru. Proposed "total quality control" in 1956.
felt: Fibrous material of interlocked fibers by mechanical, chemical, moisture, or heat action.
female mold: The mold cavity or portion of a mold which encloses the molded article.
fiber: Continuous or short discontinuous lengths of monofilamentary materials. A filament with a finite length that is at least 100 times its diameter.
fiber content: The weight or volume percent fraction of fiber in a composite.
fiber count: Count of fibers in a section of an individual ply.
fiber diameter: The dimensional measurement across the diameter of an individual filament.
fiber direction: The longitudinal axis alignment as referenced to a specific axis.
fiber efficiency: The effect on mechanical strength as a result of varying fiber length.
fiberglass: A fiber made by drawing glass.

fiberglass chopper: A mechanical device or equipment for cutting (chopping).

fiberglass reinforcement: Glass fabric, mat, roving or strand used to reinforce thermoset and thermoplastic composites.

fiber-matrix interface: The interfacial zone or region separating the fibers from the matrix materials, differing from them chemically, physically, and mechanically.

fiber orientation: Fiber alignment in a nonwoven or a mat laminate where the majority of fibers are in the same direction, resulting in a higher strength in that direction.

fiber pattern: Visible fibers on the surface of laminates or moldings; the thread size and weave of glass cloth.

fiber placement: Robotic or automated system for positioning or placement of fibrous materials into specific locations.

fiber reinforced plastic (FRP): Term to describe a fabric cloth, mat, strand, or any other fiber reinforced composite.

fiber show: The prominence of strands or bundles of fibers not covered by matrix resin system (at or above the composite surface).

fiber strain in flexure: The midspan maximum strain in the outer fiber.

fiber stress in flexure: Maximum stress of fiber in a composite beam occurs at the midpoint when the simple beam is supported at two points and loaded at the midpoint.

fiber wash: The spreading or loosening of fibers (woven or non-woven) by movement during a resin application process.

fibrillar: Molecules assembled in a fiber-like form.

Fick's equation: An equation to determine migration of moisture.

filament: A fiber of indefinite length, a continuous fiber of circular or noncircular cross section.

filaments: Individual indefinite length fibers used in roving tows or yarns.

filament weight ratio: The filament weight ratio to total weight of a composite.

filament winding: Process in which continuous reinforcement, preimpregnated or impregnated during the winding process is placed over a rotating form (mandrel) in a controlled manner. The mandrel is removed after the required layers are built up, consolidated or cured.

filamentary composite: Laminate composed of multiple layers in which continuous filaments are in nonwoven, parallel, uniaxial patterns and can be combined into specifically oriented multiaxial laminates.

fill: Fibers or yarn at right angles to the warp of a woven fabric.

filler: An inert additive substance which when added to a resin system, reduces cost. Fillers also alter or improve physical properties.

filler sheet: A pressure pad or sheet of deformable elastometric material that aids in providing uniform pressure over an area to be bonded.

fillet: The form or shape of material (rounded) filling an angle corner at the intersection of two flat surfaces.

filling yarn: Transverse fibers or threads (wept yarn) in a woven fabric that run perpendicular to the warp.

fill time: The time from start of resin injection to complete filling of the mold.

film: A term describing a very thin sheeting of a thickness (less than 10 mils, 0.010 inch).

film adhesive: Thin adhesive in the form of a thin, dry, resin film with or without a carrier.

film stacking: An early laminating method for producing fiber reinforced thermoplastics in which layers of reinforcement were interleaved with the matrix materials in film form. No longer widely practiced due to the problems in adequate wetting of the fibers.

fin: Material (flash) attached to part at parting line or vent holes.

fines: Very small particles among larger ones within a molding or other powder.

finish: A material or mixture of materials for treating glass or other fiber reinforcements. Finishes improve the resin to fiber bond as well as improve physical properties of a laminate.

fire retardant: Chemical that is used to reduce the tendency of a resin to burn or spread flame, reduce ignition and smoke emission.

first-ply-failure: Surface ply or ply group that fails in multidirectional laminates.

fish-eye: A round small mass in a coating surface, film, sheet or plastic, near the surface, caused by incomplete blending of the mass with surrounding material.

flake: Description of a material form for a dry, thin and flat particle.

flame resistance: Ability for a material to extinguish flame once the source of heat is removed. *See also* self-extinguishing resin.

flame retardant resin: Resin system compounded with certain chemicals and additive which reduce or eliminate the ability to burn.

flame retardants: Chemicals and additives that are used in a resin system to reduce or eliminate resin burning.

flame spraying: Process for applying a plastic or molten metal using a cone of flame onto a surface.

flame treating: A surface rendering process that allows lacquers, paints, inks, adhesives, etc., to bond after the surface is bathed in an open flame.

flammability: A measure of a material's extent to which a material will support combustion.

flash: Excess material that flows or is squeezed from the mold cavity during the molding process, usually along the parting line of a molded item and removed upon ejection of the item.

flash mold: Type of mold designed to allow excess molding material to escape as flash during the molding cycle.

flash point: The temperature (lowest) at which a combustible liquid will give off a flammable vapor that can burn momentarily.

flexibilizer: Additive that makes resins and resin systems more tough, flexible or impact resistant.

flexible molds: Elastomeric or rubber used to mold parts (cast, laminated, sprayed, other) with rough and undercut surfaces.

flexural modulus: In a test specimen, ratio within the elastic limit, for the applied stress on a test specimen in flexure, to the corresponding strain in the outermost fibers.

flexural rigidity (fibers): Measure of the rigidity (dyne-cm) of individual fibers or strands. The force couple required to bend a specimen to unit radius of curvature.

flexural rigidity (plate): Measure of the rigidity (in/lb) of a plate.

flexural strength: Maximum stress that the surface fibers in a beam can withstand in bending. Flexural strength is the unit resistance tensile strength of outermost fibers to the maximum load before failure by bending.

floating platen: A multi-daylight platen that can move independently when located between the upper press head and the press table.

flock: A filler or reinforcement composed of short fibers of cotton or synthetic materials, used in molding compounds.

flow: Movement of adhesive or resin system (measure of fluidity) during processing by injection, compression, or transfer molding, as well as the continuously gradual distortion of a material under continued load.

flow line: The marks or weld lines on a molded piece caused by the meeting of two resin system flow fronts during molding.

fluidized-bed coating: A powder coating technique which consists of heating an item or surface to be coated well above the melting or fusing point of the thermoplastic or thermoset material being used as the coating, dipping the object in an aerated fluidized bed of the powder coating material and upon the material sticking to the item, removing it from the bed.

fluorescent pigments: A pigment which absorbs and emits radiation of different frequencies causing a glow.

fluorinated ethylenepropylene (FEP): Member of the fluoroplastics (fluorocarbons) family of plastics.

fluoroplastics: Polyolefin polymers that possess outstanding electrical, chemical, flame and fire, weatherability, and other properties.

fluted core: Ribs of integrally woven reinforcement material used between skins of a laminate.

flux: An additive for plastics and other compositions that improves flow or fluidity.

foamed plastic: Resin systems flexible or rigid, containing cells, closed or open, interconnected, and appearing sponge-like, with a density of less that 1 to more than 40/lb per ft.

foaming agent: In resin systems, plastic and rubbers, additives and chemicals that liberate gases when part of a chemical reaction or other means of heat generation. The gases are trapped in the material and upon heating cause the material to expand and form a cellular internal structure.

foam-in-place: Dispensing foam systems at the exact location where the cavity to be filled is located (by machine or container mixing).

foam spray: Application of foam by spraying equipment. Used in machine and construction industries.

force: In compression or matched die molding, the upper moving or male half of the mold, which enters the female half exerting the pressure.

force plate: Plate used for holding male plugs in place for compression or matched die molding.

forming: Is the process in which plastic and composite materials formed by molding, extrusion or other methods, are shaped to a final configuration.

fracture: Rupture within a body caused by internal or external forces and resulting in surface or internal body separation.

fracture stress: Load at fracture or true normal stress of a test specimen.

fracture toughness: Measurement of materials strength that already contains flaws.

friction, coefficient of: *See* coefficient of friction.

frothing: Polyurethane foam application process

where blowing agents or air enter under pressure into the liquid foam mixture.

fungus resistance: Resisting degradation by fungi.

furan resins systems (Also furane): Dark-colored, thermosetting low to high viscosity polymer resin systems. They are based on furfural or furfuryl alcohol.

fuse: In plastics, a term describing a bonding or connection process.

fused coating: A thin melted and solidified coating connected to a base material.

fused quartz: *See* silica.

fusible mandrel: Former for the manufacture of hollow objects which may be melted out after solidification of the composite.

fuzz: A collection of broken filaments or fibers from roving, strands or yarns that have been passed over a particular point.

G

gage length: The original length of that portion of a tensile or compression test specimen over which strain or change in length under load is to be measured.

gap: An open space or joint in filament winding, the unintentional space between two successive windings.

gate: In resin injection and transfer molding, the opening through which molten resin enters the cavity. Sometimes the gate has the same cross section as the runner but usually it is much smaller and restricted.

gel: In thermoset resin systems, the initial jellylike solid phase developing during the formation of a resin from a liquid where there is a semisolid system consisting of a network of solid aggregates in which liquid is held.

gelation: The point in a thermoset resin system cure when the resin viscosity has increased to a gel state through the formation of infinitely large polymer networks.

gelation time: For thermosetting resins the interval of time between introduction of a catalyst into a liquid resin system and gel formation.

gel coat: A quick-setting thermoset resin system applied to the surface of a mold and gelled prior to layup. A gel coat becomes an integral part of the finished laminate, and is usually used to improve surface appearance, provide color to the laminate surface, improve chemical solvent and other surface resistance characteristics, etc.

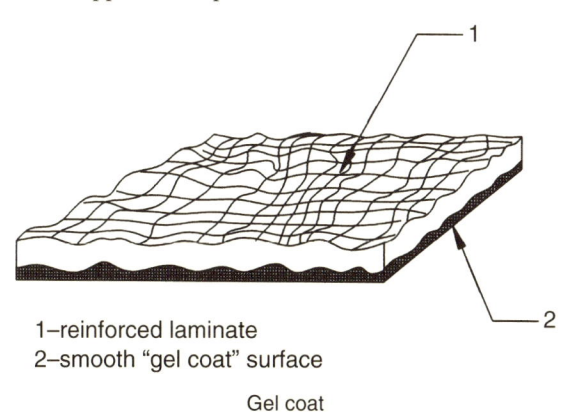

gel permeation chromatography (GPC): A form of liquid chromatography in which polymer molecules are diffused and separated by their ability or inability to penetrate the material in the separation column as they pass from the top to the bottom of the column.

gel point: The stage at which a liquid begins to exhibit increased viscosity and pseudo-elastic properties as observed from the inflection point on a viscosity-time plot.

gel time: Commonly used to describe, under defined conditions, the time in minutes after which a thermosetting resin system loses its initial liquid consistency, becomes unworkable or nonflowable.

geodesic: The exact position and shortest distance between two points on a surface.

geodesic isotensoid: In filamentary structures the constant stress in any given filament at all points in its path.

geodesic-isotensoid contour: A dome contour in filament wound reinforced plastic pressure vessels, in which the filaments are placed on geodesic paths so they will exhibit uniform tensions throughout their length under pressure loading.

geodesic ovaloid: A dome end contour where, the fibers form a geodesic line—the shortest distance between two points on a surface of revolution. The proportioned forces exerted by the filaments meet hoop and meridional stresses at any point.

glass: A hard and relatively brittle inorganic product of fusion. When cooled to a rigid condition without crystallizing it has a conchoidal fracture. Some typical composite reinforcement fibers include A glass, C glass, E glass and S glass.

glass cloth: Fiber material woven from glass.

glass fiber: A glass filament, cut to a measurable length. Spun yarns are usually made from short length staple fibers.

1–reinforced laminate
2–smooth "gel coat" surface

Gel coat

glass filament: A form of glass, drawn to a small diameter (usually less than 0.005 inch in diameter) and extreme length.

glass filament bushing: The tool or device through which molten glass is drawn when making glass filaments.

glass finish: A material applied to the surface of a glass reinforcement as an adhesion promoter to improve the bond between the glass, the plastic resin matrix and enhance the physical properties of the reinforced plastic.

glass flake: A filler for resin systems made of very thin, irregularly shaped flakes of glass formed from shattered continuous thin walled tubes of glass.

glass former: Oxide which forms into glass easily. An additive that can contribute to the network of silica glass.

glass mat: Glass fibers cut and layered into a bound or unbound thin mat.

glass spheres: Solid or hollow glass filler shaped like beads, bubbles or spheres.

glass stress: Glass fibers cut and layered into a bound or unbound thin mat. The stress calculated using the load and the cross sectional area of the reinforcement. Found in a filament wound part (pressure vessel).

glass temper: Introduce compressive surface stresses into glass by rapid cooling from above the annealing point.

glass transition temperature (T_g): An approximate midpoint of the temperature range over which the glass transition takes place. The temperature at which increased molecular mobility results in significant changes in the properties of a cured resin system. Resulting in the appearance of a rubbery flexible material (or solid to glass state).

glassiness: In a pultrusion, a glassy, marbleized, streaked appearance.

glitter: A filler made of special decorative flake materials large enough so that each separate flake produces a plainly visible sparkle or reflection.

gloss: The shine, sheen, or lustre causing a mirror-like reflection.

glue: An adhesive or term used "to connect" and originally made from a hard gelatin obtained from hides, tendons, cartilage, bones, etc., of animals.

glue line: The line or layer where adhesive attaches two adherends.

graft copolymer: A polymeric chain or polymer to which side chains of a different or same type are attached or grafted.

granular structure: The nonuniform appearance of finished plastic material due to incomplete fusion of particles or fillers within the mass or on the surface.

graphite: A crystalline allotropic form of carbon.

graphite fiber: Fiber made from a precursor by carbonization, oxidation, and graphitization process.

graphitization: The process to produce a turbostratic graphite crystal structure by pyrolyzation of an organic material at temperatures in excess of 1800°C in an inert atmosphere.

green strength: The ability or strength of a material, while not completely cured, set or sintered, to be removed from a mold and handled without distortion.

greige: Textile fabric, yarn or fiber before finishing, bleaching or dyeing. Also called gray goods, greige goods, or greige gray.

grex: The gram weight of 10 kilometers of yarn or fiber.

grind: Mechanical process for reduction of particle sizes in materials.

grit blasting: A surface treatment or finishing process in which steel, sand or other grit materials are blown on to the walls of a mold or part to produce a roughened surface.

guide pin bushing: The bushing through which guide pins move in a closing mold.

guide pins: Pins which guide mold or tool halves into alignment on closing.

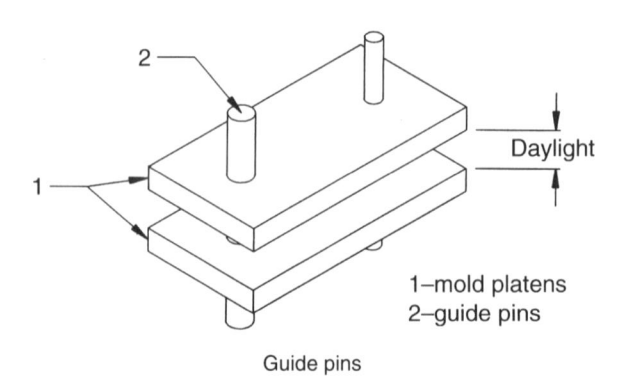

Guide pins
1—mold platens
2—guide pins

gum: Any class of colloidal and amorphous substance or mixture which, at ordinary temperatures, is either a very viscous liquid or a solid which softens gradually on heating and which either swells in water or is soluble in it.

gunk: A viscous premix of a resin system, filler, fibers, pigments, etc.

gusset: A tucked in, folded in or piece used to give added size or strength in a particular location of an object or flattened tubular film.

H

halocarbon plastics: Group of plastics composed of resins generated from the polymerization of monomers consisting of a carbon, halogen or halogens.

hand: Softness of a piece of fabric determined by touch (individual judgment).

hand layup: Laminating process of placing (and working) successive plies of reinforcing material or resin-impregnated reinforcement in position on or in a mold by hand.

hank: Looped or coiled fiber, strand, tow, etc., of definite length.

hardener: Substance, mixture or additive material used to promote or control curing action by taking part in the reaction (as opposed to a catalyst).

hardness: Resistance of a plastic to surface indentation or scratch penetration usually measured by the depth of penetration (arbitrary units) of a blunt point under a given load using a particular instrument *See also* Rockwell, shore "hardness".

harness satin: Weaving pattern of warp and tow fabric crosses producing a satin appearance.

hat: A section or structural member in the shape of a hat.

haze: Degree of cloudiness in a plastic material.

headers: Support structures bonded to a tool to increase strength.

heat activated adhesive: Dry adhesive film that becomes tacky or fluid by application of heat, or heat and pressure, to the assembly.

heat cleaned: A process in which glass or other fibers are exposed to elevated temperatures to remove preliminary sizings or binders not compatible with the resin system to be applied.

heat convertible resin: Thermosetting resin system convertible by heat to a solid.

heat distortion point: The deflection temperature at which a standard test bar deflects a specified amount under a stated load.

heat endurance: The time of heat aging that a material can withstand before failing a specific physical test.

heat resistance: Property or ability of plastics and elastomers to resist elevated temperatures.

heat sealing: Method of joining thermoplastic materials and films by simultaneous application of heat, pressure and contact.

heat sealing adhesive: Thermoplastic film adhesive that melts between adherend surfaces by application of heat.

heat sink: A material, substance or device for the absorption or dissipation of unwanted heat.

heat treat: Process of annealing, hardening, tempering, etc.

helical winding: Filament winding process in which a filament band moves along a helical path, not necessarily at a constant angle, except in the case of a cylinder.

heterogeneity: Material uniformity within a body consisting of dissimilar constituents separately identifiable.

hexa: Shortened name for hexamethylenetetramine that is used for curing novolacs.

high-frequency heating: Heating of materials by dielectric loss in a high frequency electrostatic field.

high load melt index: Rate of flow of a molten resin.

high pressure laminates: Laminates molded and cured at pressures not lower than 6.9 MPa (1.0 ksi), and more commonly in the range of 8.3 to 13.8 MPa (1.2 to 2.0 ksi).

high-pressure molding: Molding processes using pressures above 200 psi.

high-pressure spot: Area of a laminate containing very little resin.

high-speed resin transfer molding (HSRTM): Resin transfer molding (RTM) systems with a production rate of one part per minute or faster.

hob: Master model of hardened steel used to sink the shape of a mold into a soft metal block.

hobbing: Process forming multiple mold cavities by forcing a hob into soft metal cavity blanks.

hold-down groove: Small groove cut into the side wall of the mold surface to assist in holding the molded article in that mold side while the mold opens.

homogeneous: Descriptive term for a material of uniform composition throughout and whose properties are constant at every point.

homopolymer: Polymer consisting of only one monomeric species.

honeycomb: Resin impregnated paper, glass fabric, sheet material, or metal foil, formed into connected hexagonal open cells. Lightweight core material for sandwich laminates.

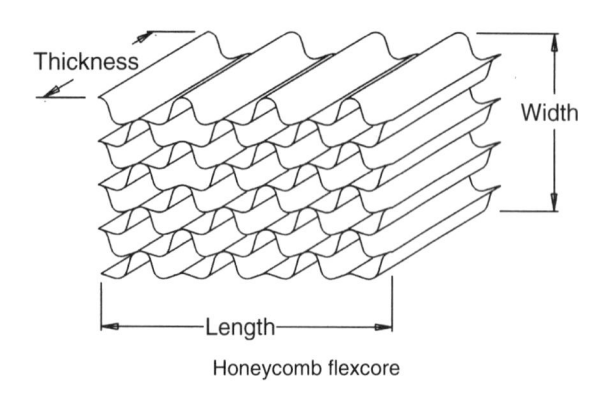

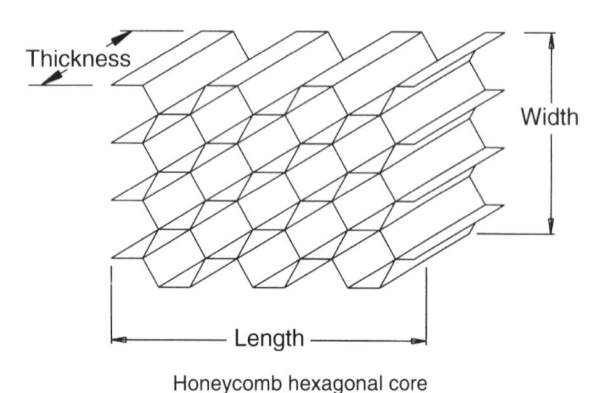

Honeycomb flexcore

Honeycomb hexagonal core

hoop fibers: Parallel fibers at near 90° to component centerline or axis of rotation.

hoop stress: Circumferential stress in a material of cylindrical form subjected to internal or external pressure.

horizontal shear strength: Three point short-beam-shear test. This interlaminar shear test is for approximate measurements.

hot bond repair: Repair made using preimpregnated fabric and adhesive, cured at elevated temperatures (usually more than 250°F).

hot gas welding: Technique for joining thermoplastic materials using a jet of hot air from a welding torch to soften and join together at the softened points.

hot isostatic pressing (HIP): Consolidation of a (metal matrix) composite preform under high temperatures and fluid pressures in a pressure vessel.

hot melt adhesive: Adhesive that is applied in a molten state and forms a bond on cooling.

hot-setting adhesive: Adhesive that requires a temperature at or above 100°C (212°F) to set.

hot-stamping: Marking plastics with heated metal dies onto the face of the plastics.

hybrid composite: A composite with two or more types of reinforcing fibers.

hydraulic press: A pressing device in which a molding force is created by the pressure exerted on a fluid.

hydrocarbon plastics: Plastics composed of resins consisting of carbon and hydrogen only.

hydroclave: Pressure vessel that uses water as the pressurizing medium instead of using a gas as in autoclaves.

hydrogenation: Chemical process whereby hydrogen is introduced into a compound.

hydrolysis: Chemical reaction of a substance with water.

hydromechanical press: A pressing device in which the molding forces are created partly by a mechanical system and partly by an hydraulic system.

hydrophilic: Capable of adsorbing or absorbing water.

hydrophobic: Capable of repelling water.

hygroscopic: Capable of adsorbing and retaining atmospheric moisture.

hygrothermal effect: A change in properties due to moisture absorption and temperature change.

hysteresis: Energy absorbed in a complete cycle of loading and unloading. The energy is converted from mechanical to frictional energy (heat).

I

ignition: The start of combustion.

ignition loss: Change in weight before and after burning

immediate set: Deformation immediately after removal of the deforming load.

immiscible: Term to describe two or more insoluble fluids incapable of mixing.

impact bar: A fracture test specimen used for shock testing.

impact resistance: Ability of a plastic or composite specimen to resist shock from a pendulum-type impact machine test.

impact shock: Stress transmitted to a surface by a sudden energy impact.

impact test: Measures the energy required to fracture a notched test bar by an impact load.

impedence: Measure of total resistance of a material (width, length, and depth) to flow of ac electricity.

impingement mixing: Injection of curing agent into resin by impacting the flow streams of the two components.

impregnate: To saturate the voids and interstices of a reinforcement with a resin system. To penetrate and soak with liquid.

impregnated Fabric: Fabric impregnated with a resin system. *See* prepreg.

inclusion: Foreign or undesirable matter found in a solid object. (particles, chips, films, etc.)

indicial notation: Sub- or superscript notation.

inert: Not chemically reactive.

inert filler: An additive for altering physical properties.

inflatable mandrel: A forming tool for filament winding, which maintains the required shape by internal pressure, and may be removed by releasing that pressure.

infrared: Electromagnetic spectrum between the visible light range and radar range where radiant heat is found for forming and curing of plastics and composites.

inhibitor: Material that retards chemical reaction or curing.

initial modulus: Initial straight portion slope of a stress-strain or load elongation curve. Also see Young's modulus.

initial strain: Strain produced in a specimen by given loading conditions before creep occurs.

initial stress: Instantaneous stress produced by force in a specimen before stress relaxation occurs.

initiator: Substance that speeds up the polymerization of a monomer and becomes a component part of the chain. Free-radical cross-linking agents.

injection mold: A two or more section mold into which a material is injected under pressure.

injection molding: Process for forming plastics by forcing the softened polymer at high pressure into a relatively cool cavity.

injection ram: Applies pressure for injection or transfer molding.

inorganic: Chemistry of all elements and compounds not classified as organic.

inorganic pigments: Natural or synthetic metallic oxides, sulfides, and other salts that impart heat and light stability, weathering resistance, color, or migration resistance to plastics.

insert: Any metal or plastic component included at or after layup or molding to allow location and connection to other subassemblies.

insert molding: Where components, such as terminals, pins, studs, and fasteners, may be molded into a part.

inspection (international quality term): Activities such as measuring, testing, gauging one or more characteristics of a product or service and comparing these with specified requirements to determine conformity.

instron: Instruments used to measure the tensile, compressive and other properties of materials (equipment of Instron Corporation).

insulation resistance: The electrical resistance of an

insulating material to a direct voltage between two conductors or systems or conductors separated only by insulating material.

insulator: A material of low thermal and electrical conductivity.

integral structure: Structure in which several structural elements, conventionally assembled together after fabrication by bonding or mechanical fasteners, are instead attached and combined in one operation.

integrally heated: Description of tooling that is self or internally heating.

intensifier: Flexible material used to apply high pressures into tight radii during bag molding operations.

interface: Boundary between physically distinguishable constituents of a composite (reinforcement to resin system).

interface barrier: A material applied to the surface of a tool to prevent contamination during cure.

interlaminar: Descriptive term pertaining to anything existing entirely within a single ply (fracture, void, etc.).

interlaminar shear strength: Maximum shear stress between layers of a laminate.

interlaminar stresses: Three stress components related to the thickness direction of a plate and significant only if the thickness is greater than 10% of the length or width of the plate.

intermingled yarn: Multifilament interlaced yarn where cohesion is imparted to the filament bundle.

internal mixers: Mixing machines containing cylindrical containers in which mixed materials get deformed by rotating blades or rotors.

interpenetrating network (IPN): The combination of an easily processed brittle thermosetting resin with a thermoplastic polymer which is more difficult to process, such that a continuous path can be traced through the material in either component. Properties are dependent on both of the constituent materials and on the phase morphology.

interphase: Boundary region area between a bulk resin or polymer and an adherend (reinforcing material).

interply: Between laminate plies.

interply hybrid: A reinforced laminate where different materials are used within a specific layer.

introfraction: Fluidity and wetting property change of an impregnating material.

invariant: Stress, strain, stiffness and compliance all have linear and quadratic invariants and constant values for all orientations of the coordinate axes.

ionic polymerization: Polymerization which takes place through ionic intermediates instead of through neutral species.

iosipescu shear test: Test named after its developer, using special fixtures to perform an composite specimen, in-plane shear strength test.

irradiation: The bombarding of polymers, plastics and composites using subatomic particles, (alpha, beta, or gamma rays) to initiate polymerization or copolymerization.

isocyanate plastics: Resin systems made by the reaction of organic isocyanates with other compounds.

isocyanate resin: One component of a resin system that contains organic isocyanate radicals reacted with polyols such as polyester or polyether.

isocyanurate resin: Formed by three isocyanate molecules that react with themselves.

isomeric: The same elements united in the same proportion by weight, but different in one or more properties due to differences in structure.

isotactic: Polymeric molecular structure (highly crystalline) containing sequences of regularly spaced asymmetric atoms that are arranged in similar configuration in the primary polymer chain.

isotensoid: Constant tension, particularly in filament winding.

isotropic: Ability to react consistently even if stress is applied in different directions not directionally dependent.

izod test: Destructive, pendulum-type single-blow impact test where a notched specimen is fixed at one end and broken by a falling pendulum.

J

jet spinning: A hot gas melt spinning process using a directed blast or jet of hot gas to pull molten polymer from a die lip and extend it into fine fibers.

jig: Tool or device for holding parts together during a fabrication operation. Also know as a fixture and can be used to hold parts together during a bonding operation.

joint: The location at which two adherends are held together with a layer of adhesive.

joint, butt: A type of edge joint in which the edge faces of the two adherends are at right angles to the other faces of the adherends.

joint, edge: The joint made by bonding the edge faces of two adherends.

joint, lap: Bonding overlapped portions of an adherend by placing one adherend partly over another.

joint, scarf: A joint made by cutting away similar segments of two adherends and bonding the adherends with the cut areas fitted together.

joint, starved: A joint that has an insufficient amount of adhesive to produce a satisfactory bond.

just-in-time (JIT): Production control techniques that minimize inventory by delivering parts and material to a manufacturing facility just before they are incorporated into a product.

jute: A fibrous, felt material obtained from plant stems used to absorb excess resin and it acts as a pneumatic bleed passage during vacuum bag molding operations.

K

kerf: Width or space of a cut made by any means.

Kevlar fiber: DuPont Company trade name for an aromatic polyamide (aramid) fiber.

K factor: Coefficient of thermal conductivity or amount of heat that passes through a unit cube of material in a given time when the difference in temperature of two opposite faces is 1°.

kick over: Shop jargon/slang expression describing the curing of a thermoset resin system.

kirksite: High thermal conductivity alloy of aluminum and zinc used for the construction of molds and tools.

kitting: Process of placing two or more plies together (hand layup) in proper orientation prior to their application to the mold or tool surface. Helps to improve both productivity and producibility.

knife coating: Method of coating fabric, paper or other substrate in which the substrate, in the form of a continuous moving web, is coated with a material whose thickness is controlled by an adjustable knife or bar set at a suitable angle to the substrate.

knit lines: A flaw on a molded plastic article caused by the meeting of two flow fronts during the molding operation.

knitted fabric: Fabrics produced by interlooping chains of yarn.

knockout pin: An ejector pin or device for pushing out a cured piece from a mold.

knuckle: The point at the end of a way-wound roving ball where the roving reverses its axial direction.

knuckle area: The y-joint or area of transition between sections of different geometry in a filament-wound part.

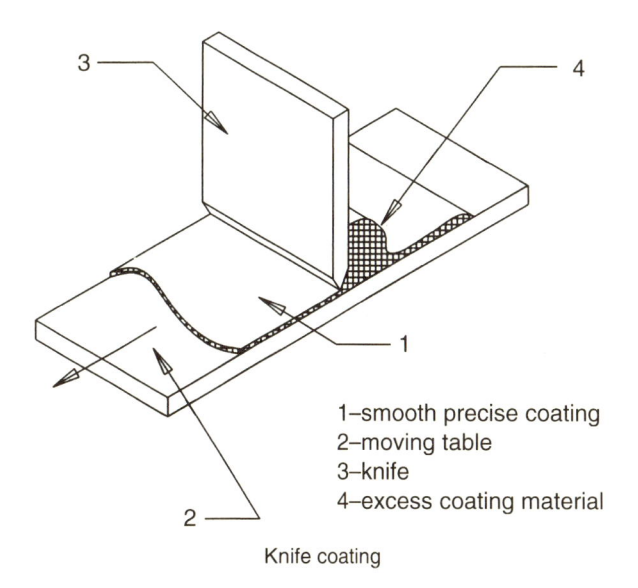

1–smooth precise coating
2–moving table
3–knife
4–excess coating material

Knife coating

L

lacquer: A protective coating made from a solution of natural or synthetic resin, in an evaporating solvent.

lamina: A flat or waved ply or layer of unidirectional composite or fabric material.

laminae: Plural of lamina. Materials used to form laminar composites.

laminate: The act of laying up successive plies of material. To unite laminae with a bonding material (matrix resin system).

laminate coordinates: A system to describe the properties of a laminate.

laminate family: Separate laminates sharing the same number of plies, angles and configurations.

laminate orientation: The exact configuration to identify the sequence, angles, position and other characteristics of a cross-plied composite laminate.

laminate ply: One layer of fiber/resin system bonded to adjacent layers.

laminate thickness: Thickness of single or multiple plied material.

laminated, cross: Laminate where some layers of material are oriented at right angles to the remaining layers.

laminated, parallel: All layers of material are oriented approximately parallel.

laminated plastics: Laminated structural shapes that include angles, plates, sheets, rods, tubes, channels and other configurations.

laminated plate theory: The lamination theory and theoretical basis used for the analysis of composite laminates where each ply or ply group is treated as a quasi-homogeneous material.

lamination: The process of applying layers to form a laminate.

land: The portion of the die which limits closure such as the horizontal bearing surface of a mold by which excess material escapes.

land area: Surface areas of a mold which contact each other when the mold is closed.

lap: The amount of overlay between successive (filament) windings, intended to minimize gaping.

lap joint: Joint formed by bonding overlapped portions of two adherends.

latex: Generic term for any stable dispersion of insoluble resin particles in a water system.

lattice pattern: A filament winding pattern with open voids (basketweave).

lay: Orientation of filament wound ribbon, usually referenced to the axis of rotation.

lay-flat: A nonwarping property of laminating adhesives.

layup: The sequential placing of reinforcements onto/into the mold, using either wet laminating or prepreg systems.

layup mandrel: Form, mold or fixture used for shaping a part.

leach: Extracting a soluble component from a mixture by the process of percolation.

legging: Forming strings when adhesive-bonded substrates are separated.

lengthwise direction: Lengthwise is the direction that is stronger. It refers to the cutting of specimens and to the application of loads. Lengthwise is the direction of the long axis in a rod or tube.

Lexan: Trade name of a thermoplastic material.

light-fastness: Resistance to light.

light-resistance: Ability of a plastic material to resist fading after exposure to ultraviolet light or sunlight.

lignin plastics: Composed of resin systems formulated from lignin treated with heat or by reaction with chemicals.

linear expansion (coefficient of thermal expansion): Increase of a dimension, measured by the expansion or contraction of a specimen or component subjected to a changing temperature.

linear relation: A straight line relation between input and output variables.

line pressure: Air or hydraulic system operating pressure.

liner: A continuous coating or metallic bladder on the inside surface of a filament wound vessel.

liquid crystal polymer: High orientation thermoplastic polymer that is newer, melt processable when molding and has high tensile strength and temperature capability.

liquid metal infiltration: Process for immersion of metal fibers in a molten metal bath to achieve a metal matrix composite (graphite fibers in molten aluminum).

liquid resin: Organic polymeric liquid which becomes a solid upon conversion.

liquid shim: Material (epoxy adhesive) used to position components in an assembly where dimensional alignment is critical.

load: Force such as in-plane or flexural stress.

load-deflection curve: Curve in which increasing flexural loads are plotted on the ordinate axis and the deflections caused by those loads are plotted on the abscissa axis.

loading rate: Change in load per unit time.

loading space: Space in a compression mold or transfer mold pot where molding material is stored before it is compressed.

long fiber: Fiber of finite length greater than the critical length.

longitudinal modulus: The elastic constant along the fiber direction in an unidirectional composite.

longos: Low angle helical or longitudinal filament windings.

loop tenacity: Loop strength obtained by pulling two loops of the same fiber against each other to demonstrate the resistance to cutting or crushing.

loss angle: Antitangent of the electrical dissipation factor.

loss, dielectric: *See* dielectric loss.

loss factor: Product of the dissipation factor and the dielectric constant of a dielectric material.

loss index: The measure of dielectric loss. The product of the power factor and dielectric constant.

loss modulus: Term describing the dissipation of energy into heat when a material is deformed.

loss on ignition: Loss of weight (percent of total) after burning off of an organic resin from fiber laminates or an organic sizing from glass fibers.

loss tangent: *See* dissipation factor.

lost core: *See* fusible mandrel.

lot: A specific amount of material produced at one time.

low-pressure laminates: Laminates molded and cured using pressures from just contact of the plies up to 400 psi (2,800 kPa).

low pressure molding: Molding or laminating between 1–14 atmospheres.

lubricant: A component of most fiber sizes that improves the handling and processing properties of the strand.

Lucite: A trade name generically used for polymers made from methyl methacrylate.

luminescent pigments: Pigments that produce a glow effect in the dark.

lyophillic: Affinity for the dispersing medium in a dispersion.

lyophobic: In dispersions, not having affinity for the dispersing medium.

M

macerate: Process of shredding or chopping materials. To chop or shred fabric for use as filler.
machine tooling: Any mold manufactured by cutting the tool face from a solid block.
macro: Denotes the gross properties of a composite as a structural element.
macromechanics: Structural behavior of composite laminates using the laminated plate theory. Where the fiber and matrix on each ply level act as one.
macromolecules: Very large molecules of high polymers.
male mold: The protruding portion of a mold which is covered by the molded article, or a mold itself.
mandrel: The core tool around which parts are molded or from which mold surfaces are formed.
manifold: A tubing or piping system that divides inlets and outlets in multiples.
markoff: A mark, indentation or imprint in the surface of a molded part made by any cause.
mass: The unit measure for quantity of matter.
mass stress (Fibers): Grams per denier or force per unit mass per unit length.
masterbatch: A compound of plastics having a high concentration of additive materials.
master transfer gage: A female, (reversal) part taken off a master model for use as a master for making male production parts.
mat: A planar assembly or randomly oriented chopped fibers, or swirled continuous strands, of reinforcement with a binder, available in blankets of various widths, weights, and lengths.
mat binder: Dry solid or liquid material applied to fibrous mat to hold the fibers in place and maintain the shape of the mat.
matched metal molds: Metal male and female molds, usually with heat and pressure, to form reinforced plastics parts.
matrix: A thermoplastic, thermoset, glass, ceramic or metal material in which the fiber of a composite is imbedded.
matte finish: Dull, totally diffused nonreflective surface finish.
maturation: The increase and leveling off of the viscosity of sheet molding compounds.
maximum strain: Failure based on maximum strains.
maximum stress: Failure based on maximum stresses.
mean strain: Analogous to mean stress.
mean stress: Algebraic average of the minimum and maximum stresses in a cycle of fatigue loading.
mean stress: A dynamic fatigue parameter. The algebraic mean of the maximum and minimum stress in one cycle: $\sigma_2 = 1/2 (\sigma_1 + \sigma_2)$ where σ_1 is maximum stress and σ_2 is minimum stress.
mechanical adhesion: Interlocking adhesion action between adherends.
mechanical properties: The individual relationship between stress and strain. Properties of a material, such as compressive and tensile strengths, and modulus.
melamine plastics: Plastics based on melamine resins.
melt: A heated and molten metal or plastic.
melt impregnation: Process of thoroughly wetting out reinforcement fibers with molten polymer.
melt index: The amount, in grams, of a thermoplastic resin which can be forced through a 0.0825 inch orifice when subjected to 2160 gms. force in 10 minutes at 190°C.
melt strength: Strength of the plastic while in a molten state.
meniscus: Top concave surface line of a liquid (except mercury convex meniscus) when enclosed in a container such as a can, test tube, etc.
mer: Repeating structural unit of any polymer.
mesophase: Intermediate phase in the formation of carbon from a pitch precursor.
metal core: Aluminum honeycomb.

metallic fiber: Manufactured plastic-coated metal, metal-coated plastic fiber.

metallic tooling: Any mold manufactured from a metal block or plating.

metallizing: Process (vapor or chemical deposition) to apply a thin coating of metal to a nonmetallic surface.

metastable: In a plastic, the unstable condition resulting from changes in physical properties not caused by changes in composition or environment.

M glass: High (110 GPa) elastic modulus glass fiber (principally 54% silica, 13% calcium oxide) with high (8%) beryllia content.

mic-mac: Integrating micromechanics and macromechanics for the design of composites.

microcracking: Small cracks or crazes in the matrix material formed when thermal or mechanical stresses exceed the strength of the matrix.

micromechanics: For composites, the calculation of the effective ply properties as functions of the fiber and matrix properties.

micron: Unit of length replaced by the micrometer (10^{-6}m = 10^{-3}mm=0.00003937).

microspheres: Small hollow bubbles or spheres used as lightweight plastic fillers to produce (low density syntactic foam) compounds.

microstructure: A material's grain structure as seen through a microscope.

midplane: Surface layer located half-way at the midpoint through the thickness of a laminate.

migration: Extraction or movement of an ingredient from a material by contact with another material.

mil: A unit of measurement (1 mil = 0.001 in.)

milled fiber: Hammer milled very short glass fibers.

mix: Thorough combination of two or more materials.

mixing head: A mechanical device on a machine where compound components are combined.

mixing ratio: The proportions, usually measured in parts by weight, of a base material and the corresponding hardener.

mock leno weave: Woven open-mesh type fabric resulting primarily from the weave interlacing sequence.

mock-up: Full scale precise and dimensionally detailed representation of a proposed object or system.

modify: Add ingredients, fillers, pigments or other additives, to vary the physical properties of a material.

modifier: The chemically inert ingredient that is added to a formulation that changes its properties.

modulus: Number showing the measure of some property of a material (modulus, etc.).

modulus, initial: The slope of the initial straight portion of a stress strain or load-elongation curve.

modulus of elasticity: Stress/strain ratio in a plastic material that is elastically deformed.

modulus of elasticity in torsion: The ratio of the torsion stress to the strain in the material over the range for which this value is constant.

modulus of resilience: The energy that can be absorbed per unit volume without creating a permanent distortion.

modulus of rigidity: Ratio of stress to strain within the elastic region for shear or torsional stress.

modulus of rupture: The force necessary to break a specimen of specified dimensions as expressed in PSI.

Mohr's circle: Graphical representation of the variation of the stress and strain components resulting from rotating coordinate axes.

mohs hardness: A measure of the scratch resistance of a material (diamond is 10) by comparison.

moisture absorption: Pickup of water vapor from air (not water absorption) by a material.

moisture content: Amount of moisture in a material determined under prescribed conditions and expressed as a percentage of the mass of the moist specimen.

moisture equilibrium: Condition reached by a material when it no longer takes up or gives moisture from, or to the surrounding environment.

moisture vapor transmission: Rate at which water vapor passes through a material at a specified temperature and relative humidity.

mold: Cavity in which a composite part is shaped. The process of shaping.

molded edge: Edge that is molded finished and not physically altered during or after the molding cycle.

molded net: Description of a molded part that requires no additional processing.

molding: Forming a composite or polymer system into a solid specified mass.

molding compounds: Plastic systems in a wide range of forms (specially granules or pellets) to meet specific processing requirements, also BMC, DMC, SMC, TMC, XMC.

molding cycle: Period of time required for the complete sequence of operations on molding equipment to produce one set of parts.

molding material: A plastic material in varying stages of pellets or granulation, and consisting of resin, filler, pigments, reinforcements, plasticizers, and other ingredients.

molding pressure: Pressure applied directly or indirectly on the compound being molded.

molding shrinkage: Difference in dimensions between a molding and the mold cavity in which it was molded.

mold-release: Lubricant, liquid, or powder used to prevent sticking of molded articles in the mold.

mold seam: Line on a molded or laminated piece caused by the parting line of the mold.

mold shrinkage: Incremental difference between the dimensions of the molding and the mold from which it was made, expressed as a percentage of the mold dimensions.

mold surface: Side of a molded part that faced the mold (tool).

molecular weight: Sum of the atomic weights of all the atoms in a molecule.

moment: A stress couple that causes a plate to bend or twist.

monofilament: A single filament or fiber of indefinite length, strong enough to function as a yarn in commercial textile operations.

monolayer: Basic laminate unit from which crossplies or other laminate types are constructed.

monomer: A simple compound that can react with like or unlike molecules to form a polymer (smallest repeating structure of a polymer [mer]).

Mooney equation: Relates the viscosity of a concentrated suspension to the particle concentration and shape.

morphology: Overall form of a polymer structure, that is, crystallinity, branching, molecular weight, etc.

mucilage: Adhesive prepared from a gum and water.

multicircuit winding: Filament winding which requires more than one circuit before the band repeats by lying adjacent to the first band.

multidaylight: Description of a press with more than one opening.

multidirectional: Having multiple ply orientations in a laminate.

multifilament yarn: Large number of fine, continuous filaments with some twist in the yarn to facilitate handling.

multiple loads: Stress applied to a body in many directions.

N

NDI: Nondestructive Inspection process or procedure for determining the quality or characteristics of a material, part or assembly without permanently altering its properties (or characteristics).

neat resin: An unfilled, non-reinforced resin system that is plain, without any additives whatsoever.

necking: The localized reduction in cross section area of a specimen that may occur when a material is under tensile stress deformation.

needled mat: A mechanically formed mat of roving strands cut to a short length, then felted together to a substrate in a needle loom.

nesting, laminate: In reinforced composites, a laminate where the plies are placed and the yarns of one ply lie in the valleys between the yarns of the adjacent ply, i.e., nested cloth.

net molded edge: A molded part edge which is not physically altered and is in final form for use directly after molding.

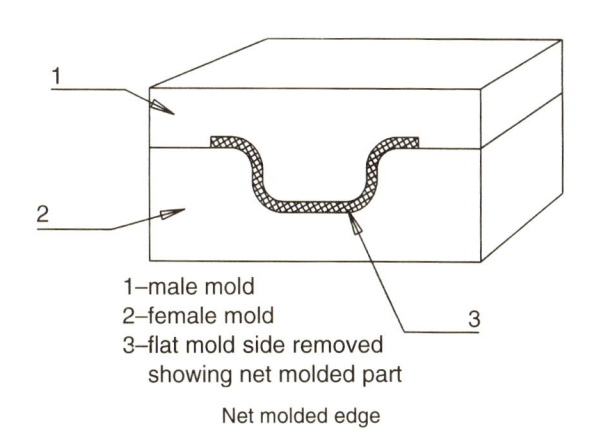

1—male mold
2—female mold
3—flat mold side removed showing net molded part

Net molded edge

netting analysis: A filament-wound structure analysis. Induced stresses in the structure are carried entirely by the reinforcing filaments, the filaments possess no bending or shearing stiffness, carry only the axial tensile loads and the resin strength is disregarded, i.e., fibers with a matrix.

Newtonian liquid: A liquid having no yield value, viscosity independent of shear rate, but one in which the rate of flow is directly proportional to the force applied.

nick: Edge or surface dent, notch or cut in a part or formed material form.

nole ring: A filament or tape wound, parallel direction, hoop test specimen. Used for measuring tension, compression and other mechanical properties by testing the entire ring or segment of it. Developed by Naval Ordinance Laboratory (NOL).

nomex: The name for a nylon fiber, paper treated, honeycomb core material bonded with thermosetting phenolic resin system that exhibits fire resistance, good formability and high strength at low densities.

nominal cured thickness: The thickness of a laminate, either prepreg or wet layup, following the cure.

nominal stress: The compression, tension or shear stress at a point calculated on the net cross section without concern for local presence of grooves, holes or other discontinuities.

nondestructive evaluation (NDE): Also referred to as nondestructive inspection (NDI).

nondestructive inspection (NDI): Inspection process or procedure using radio graphic, ultrasonic or other method to determine characteristics and quality of a material, part, subassembly or assembly without deformation or destruction during testing.

nondestructive testing (NDT): See Nondestructive Inspection.

nonhygroscopic: Will not absorb or retain appreciable amounts of moisture (water vapor) from the air.

nonmechanical stress: As in a thermoset material, caused by curing or cross-linking.

nonpolar: Does not possess a concentration of electrical charges on a molecular scale and is incapable of significant dielectric loss. Polystyrene is a nonpolar resin.

nonpolar solvent: Nonelectrically conductive liquid that can dissolve nonpolar hydrocarbon and resin compounds.

nonrigid plastic: One that has a flexural or tensile modulus of elasticity under 10,000 psi (700 Kg/cm^2) at 23°C, when determined in accordance with ASTM D 747.

nonwoven fabric: A planar random or unidirectional textile structure produced by loosely compressing together fibers, yarns, or rovings. May or may not be resin/impregnated.

nonwoven mat: A random length or direction fiber structure loosely gathered and held together using a sizing or binder.

nonwoven roving: A continuous fiber roving reinforcement made from loosely gathered textile material.

normal stress: Results from stress perpendicular to a plane upon which the forces are acting.

noryl: A trade name for a modified thermoplastic.

notch factor: Ratio of the resilience comparison of notched to un-notched specimen.

notch rupture strength: Applied load ratio of original cross section area in a stress-rupture test of a notched specimen.

notched specimen: A test specimen prepared by cutting or notching a V-shape or other shape groove along it's side to direct the point of failure.

notch sensitivity: Low notch sensitivity is found in ductile materials, high notch sensitivity in brittle materials and they both are increased by a notch, a crack, scratch or surface flaw purposely introduced into the specimen.

notch tensile strength: The maximum strength of a notched specimen tested under tensile load.

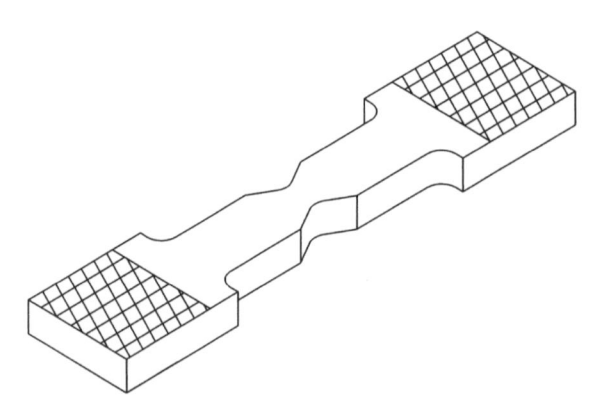

Notched specimen (tensile test)

Novolac (Novolak): A thermoplastic B-staged phenolic resin which can react with other cross-linking groups to form a thermoset phenolic.

numerical control (N/C): Control automatically electro-mechanical devices by means of digital inputs and electronic controllers.

nylon (Polyamide): Generic engineering resin based on a long-chain synethetic polymeric amide that has recurring amide groups as an integral part of the main polymer chain. Great toughness, elasticity, low coefficient of friction, good electrical properties, but dimensional stability is poorer than that of most other engineering plastics.

O

observed failure rate: For a stated period in the life of an item, the ratio of the total number of failures in a sample to the cumulative observed time on that sample. The observed failure rate is to be associated with particular and stated time intervals (or summation of intervals) in the life of the item, and with stated conditions.

observed mean life: The mean value of the lengths of observed times of failure of all items in a sample under stated conditions.

observed reliability of nonrepaired items: For a stated period of time, the ratio of the number of items that performed their functions satisfactorily at the end of the period to the total number of items in the sample at the beginning of the period.

occluded containment: Totally contained inclusion or containment in an insulation material.

off-axis: An angle not coincident with the symmetry axis. Also called off-angle.

offset modulus: Ratio for offset yield stress to the extension at an offset point.

offset yield strength: The stress (force per unit area) at which the strain exceeds an extension of the initial proportional portion (the offset) of the stress-strain curve by a specific amount.

oil resistance: A material's ability to withstand contact with oil and not deteriorate the mechanical, physical or other properties and/or cause geometric change.

oil-soluble resin: A resin that can dissolve in, disperse, or react with drying oils in a homogeneous way.

olefin: Or polyolefins are a group of unsaturated hydrocarbons by the formula C_nH_{2n}. They are similar to paraffins through the addition of root extensions such as "ene" or "ylene," as an example, petene, propylene, ethylene.

oleo resins: Semisolid mixtures of plant essential oils and resins that can consist of synthetic, material resins.

on-axis: Or on-angle and coincident with the symmetry axis.

one-shot molding: Such as a urethane foam where the system of an isocyanate, polyol, catalyst and other constituents when mixed together directly, immediately convert to foam (as distinguished from prepolymer).

opacity: Opposite of transparency. A material's ability to partially obscure a graphic image through and beneath the material surface.

opaquer: A material that, after addition to the resin system, causes a surface to which the system is applied to be opaque so that the surface beneath cannot be seen.

open cell foamed plastic: A cellular plastic where the interconnecting cells greatly exceed closed cell content. Open connecting cells allow intercell gas flow.

open molding: Process for forming parts on a single-sided male or female mold.

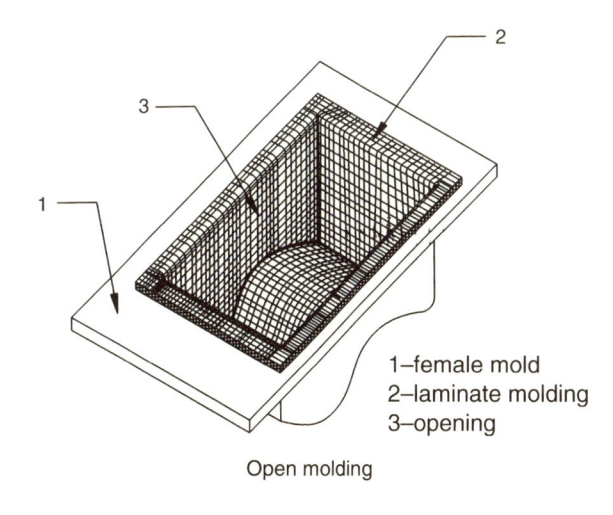

1–female mold
2–laminate molding
3–opening

Open molding

open time: The maximum duration from the application of an adhesive, prepreg or resin system to the formation of a satisfactory interlaminar bond. *See also* Working Time

operating time: The period of time during which an item performs its intended function.

operational cycle: A repeatable sequence of functional stresses.

operational requirements: All the functional and performance requirements of a product.

optimum condition: The ideal situation: the best possible condition.

optimum laminate: The greatest strength and stiffness attainable per unit cost or mass.

orange peel: An uneven rough surface like that of an orange peel. The condition can be intentional, but improper coating or molding methods could cause such an problem.

organic: Matter originating in plant or animal life or composed of carbon compounds, not including carbonates or the oxides of carbon, such as the inorganic compound, carbon dioxide. A resin system would be organic as opposed to fiberglass reinforcement. Chemical compound also containing hydrogen with or without nitrogen or other elements.

organic contamination: Contamination derived from an organic substance.

organic pigments: Compound additives characterized by good brilliance.

organosol: A PVC or other liquid phase resin dispersion containing one or more organic solvents.

orientation: Alignment of the crystalline structure in thermoplastic polymeric materials in order to produce a highly aligned structure. Oriented materials that are anisotropic and can be divided into two classes: uniaxial (one direction) and biaxial (two directions).

original inspection: The first inspection of an item or lot/batch as distinct from an inspection resulting from a prior rejection.

orthotropic: Having three mutually perpendicular planes of elasticity symmetry such as unidirectional plies, fabric, cross and angle-ply laminates.

outer skin: In a laminate layup, into an open mold, it is that side of the laminate that is against the mold.

outgassing: Evolution of gas from a material in or out of a vacuum or heat, or both.

out-of-round: Uniform radius or diameter.

out time: The time duration that a prepreg, adhesive or other compound retains its physical and mechanical properties when exposed to room moisture and temperature ambient. Adverse effects of out time include reduction of tack if a prepreg, increase in viscosity, if a pourable material, absorption of moisture and other performance characteristic reductions.

ovaloid: For the formation of the end closure on a filament wound cylinder, it is the surface of revolution symmetrical about the polar axis.

oven dry: The material condition after being heated at temperature and humidity until there is no further significant mass change.

overcoat: A thin film of insulating or coating material that is applied over a surface for the purpose of mechanical and contamination protection.

overcured: Thermoset resin system, thermal degradation and decomposition due to excessive high curing or reacting temperature or long molding time cycle.

overflow groove: Small groove or vent used in a split compression, RTM or other mold to allow resinous material to flow freely inside preventing weld or other undesirable mold lines. Also allows removal of excess resin materials from the mold.

overlap: A simple adhesive joint in a bond line where one surface or side of the adherend extends past the leading edge of another.

overlay sheet: The last surfacing "veil" or top layer of non woven synthetic or glass fiber mat applied in a cloth or mat lay up. To reduce or hide the "print through" of the fibrous reinforcement below and provide a smooth molded finish.

overtravel: The distance a carriage or eye travels beyond the ends of a part mandrel that is necessary to provide lay down of the fiber on the mandrel.

oxidation: A addition of oxygen to a compound or removal of hydrogen resulting in a reaction in which electrons are transferred. In the carbon/graphite fiber process it is the step of reacting the precursor polymer (rayon, PAN, or pitch) with oxygen.

oxirane: A three membered ring of two carbon atoms and one oxygen atom.

P

package: Method for supply of yarn, roving, and so forth in the form of units capable of being unwound and suitable for handling, storing, shipping, and use.

packaging density: The relative quantity of functions (fillers, components, interconnection devices, mechanical devices, etc.) per unit volume. (This is usually expressed by qualitative terms such as high, medium, and low.)

packing: Filling of a mold cavity or cavities as full as possible without causing undue stress or distortion of the molds or causing flash along molded edges of item.

pallet: Usually refers to either 36 or 48 cartons of fiber roving placed on a wooden or plastic skid and strapped together for shipping as a unit.

paneling: Distortion or change in shape of a container ("oil can") during storage caused by the development of reduced pressure inside the container.

parallel axis theorem: Formula for elastic moduli relative to a displaced reference coordinate frame; analogous to that for moments of inertia.

parallel laminate: A laminate of woven fabric in which the plies are aligned in the same position as originally aligned in the fabric roll. A series of flat or curved cloth-resin layers stacked uniformly on top of each other.

parameter: An arbitrary constant, as distinguished from a fixed or absolute constant. Any desired numerical value may be given as a parameter.

paraplast: A plaster-like molding material cast to the inside dimensions of a part. Part is then laid up around it and cured. After curing, "paraplast" material is washed away from inside of the part employing pressurized water.

parison: In blow molding, the hollow plastic melt tube from which a container is blow expanded and molded.

particle size distribution: Percentage, by weight or by number, of each fraction into which an amount of material (usually powder), has been classified with respect to sieve number or particle size.

particulate composite: Metallic or nonmetallic material consisting of one or more particulate constituents suspended in a matrix of another material. These particles are either metallic or nonmetallic.

parting agent: A release lubricant, wax, silicone oil, used to coat a mold cavity or surface to prevent the molded piece from sticking to it (release agent).

parting line: The mark or line on a molded piece where the sections of a mold have met prior to molding.

passivation: Reducing the chemical activity of a metal surface thereby forming an insulating layer to protect the surface.

paste: A plastic-type consistency adhesive composition that has a high order of yield value.

paste adhesive: Usually 2 or 3 part adhesive mixes, applied by brush or spatula, air or heat cured.

pay-off: Discharge fiber tow or yarn from a package or spool.

peel: A typical bond failure characterized by a layer dis-bonding from another.

peel ply: Sacrificial open-weave fiberglass or heat set nylon ply; usually the last ply placed on layup before curing. When removed from laminate, it provides a continuous rough and clean surface for subsequent bonding operations eliminating bag wrinkles.

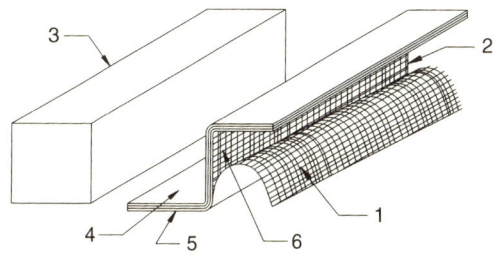

1–peel ply being removed
2–rough fabric pattern surface ready for bonding or finishing
3–master pattern
4–smooth side front of mold
5–laminate mold copy
6–rear of mold

Peel ply

peel strength: The force per unit width, (usually pounds per inch of width) required to peel one sheet of material from a base material.

penetration: The entering of an adhesive into an adherent, measured by the depth of penetration of the adhesive into the adherend.

peptides: Low molecular weight polymers of a-amino acids, having molecular weights under 10,000. Higher molecular-weight species are called polypeptides or proteins.

perforated release film: Film pierced or perforated to control resin bleed and air removal from the laminate during compaction and consolidation.

permanence: The property of a plastic or adhesive bond which describes its resistance to deteriorating and appreciable changes in characteristics with time and environment.

permanent set: The deformation (such as creep) remaining after a specimen has been stressed in tension or compression for a definite period and released for a definite period.

permeability: The passage, rate of passage or diffusion of a gas, vapor, liquid, or solid through a barrier or material without physically or chemically affecting it.

permittivity: *See* dielectric constant.

pH: Measure of acidity or alkalinity, the negative log of hydrogen ion concentration expressing the degree of acidity or alkalinity of a substance. A neutral substance is 7.0, acid under 7.0 and alkaline solution over 7.0.

phase angle: *See* dielectric phase angle.

phenolic (phenolic resin): A thermosetting resin or compound produced by the condensation of an aromatic alcohol with an aldehyde, particularly of phenol with formaldhyde.

phenylene oxide: Resin systems, when molded result in excellent dimensional stability, very low moisture absorption, superior mechanical and electrical properties over a wide temperature range. Resists most chemicals but is attacked by some hydrocarbons.

phenylsilane resins: Thermosetting copolymers in solution of silicone and phenolic resins.

photoelasticity: When subjected to stress, the changes in the optical properties of isotropic, transparent dielectrics.

photomaster: An artwork master.

photoresist: A material, sensitive to portions of the light spectrum. When properly exposed can mask portions of a base material with a high degree of integrity.

photoresist image: Exposed and developed image in a coating on a base material.

phthalate esters: Plasticizers produced by the direct action of alcohol or phthalic anhydride. Most widely used of all plasticizers.

physical catalyst: Radiant energy that can promote or modify a chemical reaction.

physical properties: Other than mechanical properties, that pertain to the physics of a material; for example, density, electrical conductivity, heat conductivity, and thermal expansion.

pick: Filling yarn that runs crosswise to the entire width of a woven fabric at right angles to the warp, fill, woof or weft.

pick count: Number of filling yarns per inch of woven fabric.

pick-up roll: A spreading device roll for picking up the adhesive or resin in a reservoir of the material.

pill: Preformed molding compound or material.

pinch-off: Raised edge around the cavity of a blow molding mold that seals off the part and separates excess material as the mold closes around the parison.

pinhole: A tiny hole or holes in the tool side surface of, or through, a plastic molded or coated item caused by solvent, by-product, reactant or other particles at the surface material.

pit: A surface imperfection, in the form of a small hole or crater.

pitch: A high molecular weight residual petroleum (crude oil) product used in the manufacture of certain carbon fibers.

plain weave: A simple fabric weave pattern or configuration whereby each warp end goes over one fill and under the next, and whereby each fill goes over one warp end and under the next.

planar: Lying essentially in a single plane.

planar helix winding: A winding in which the filament path on each dome lies on a plane which intersects the dome, while a helical path over the cylindrical section is connected to the dome paths.

planar winding: Filament winding where the filament path lies on a plane intersecting the winding surface.

plane strain: Two-dimensional simplification for stress analysis, applicable to the cross section of long cylinders.

plane stress: Two-dimensional simplification for stress analysis that is applicable to thin homogeneous and laminated plates.

plastic: A material from any one of a large and varied group consisting of or containing as an essential ingredient, an organic substance of large molecular weight. While solid in the finished state, at some stage in its manufacture the material has been molded or can be cast, formed, extruded, etc., into various shapes.

Usually through the application of heat, pressure, or a combination thereof causing material flow.

plasticate: To soften by heating, mechanical means or kneading (usually thermoplastic).

plastic deformation: A change in shape, dimensions or deformation of an object that remains permanent after removal of the load causing it ("plastic flow").

plastic flow: Is a form of plastic deformation caused under the action of a sustained force (semisolids flow during the molding of plastics).

plasticity: A property that allows a material to be deformed continuously and permanently without rupture after the application of a force that exceeds the material yield value.

plasticize: To soften a material and make it more flexible or moldable either by adding a plasticizer or applying heat.

plasticizer: A material added to a resin system, polymer or plastic to increase its flexibility, workability, or distensibility. The addition of a plastizer can cause a reduction in melt temperature, visocity, or lower the elastic modulus of the solidified resin.

plastic memory: Ability of a thermoplastic material, stretched while hot, to return to its unstretched shape upon being reheated.

plasticorder: A laboratory device (plastigraph) used to predict the performance of a plastic material by measurement of viscosity, temperature, and shear rate relationships.

plastics tool: A mold or shaping surface constructed of thermoset reinforced laminates or mass cast materials.

plastometer: Instrument for determining thermoplastic resin flow properties. Molten resin is forced through a special die or orifice at a specific temperature and pressure.

platen: A mounting plate or plates of a press to which a mold or tool assembly is fastened.

plied yarn: Yarn made by collecting two or more single twisted or untwisted yarns.

plug: Male shape or form identical to the finished object, over which a female mold is formed.

ply: A lamination layer of reinforcing material (fabric, mat, etc.) in a composite part or layup. A number of plies used will determine the part thickness.

ply drop: Laminate thickness reduction in areas by eliminating or removing plies.

ply group: A group of continuous plies with the same angle.

ply on ply: Method of laying up composite plies directly one on another onto the tool. Eliminates initially placing the material on mylar layup templates (LTs) and subsequent transfer to the tool.

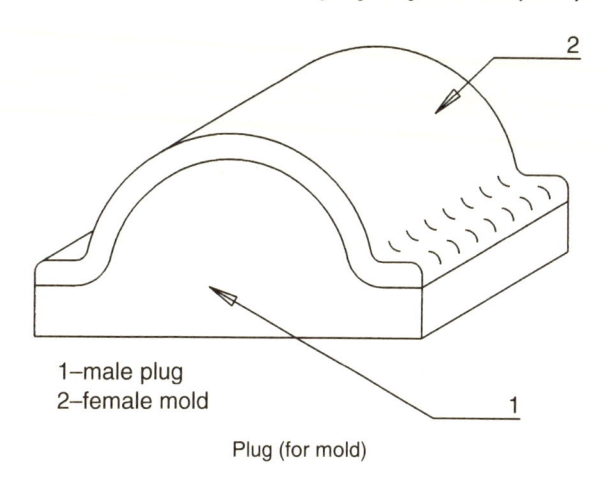

1–male plug
2–female mold

Plug (for mold)

ply strain: Apply to those parts of a ply which, by the laminated plate theory, are the same as those of the laminate.

ply stress: Apply to those parts of a ply, which vary depending on the materials and angles in the laminate.

plywood: Cross-bonded assembly of layers of veneer. Veneer in combination with a lumber core or plies joined with an adhesive.

poise: Unit measure of viscosity. (100 poises = 1 centipoise) Shear stress is expressed in dynes per square centimeter to produce a velocity gradient of one centimeter per second per centimeter.

Poisson's ratio (PR): As a material is stretched, the cross-section area changes as well as its length. Poisson's ratio is the constant relating these changes in dimensions.

Equation:

$$PR = \frac{change \; \epsilon \; width \, per \, unit \, width}{change \; \epsilon \; length \, per \, unit \, length} = \frac{\Delta C/C}{\Delta L/L}$$

i.e.:
  PR = 1/2 for rubbery materials
  PR = 1/4 to 1/2 for glass or crystal

polar matter: Substance that can dissolve in water and hydrophillic solvents.

polar winding: For filament winding where the filament path passes on a tangent to the polar opening at one end of the chamber and tangent to the opposite side of the polar opening at the other end.

polyacrylate: Thermoplastic resin system made by polymerization of an acrylic compound such as methyl methacrylate.

polyacrylonitrile (PAN): A polymer used as the base material or precursor in the manufacture of certain carbon fibers.

polyamide: A polymer in which the structural units are linked by amide or thioamide groupings (repeated nitrogen and hydrogen groupings). Many polyamides such as nylon are fiber forming.

polymide curing agents: These agents or hardeners were epoxide resins that contain free amino groups.

polyamideimide: Polymers containing both amide (nylon) and imide (polyimide) groups used for laminating and other molding.

polyarylsulfone (PAS): A thermoplastic low and high temperature resistant engineering resin with good impact, chemical, solvent and electric insulating properties.

polybends: An alloy, mixture or combination of two or more polymers, e.g., rubber and epoxy.

polybenzimidazole (PBI): A condensation polymer of diphenyl isophthalate and 3,3-diaminobenzidine. Very high-temperature and chemical resistant. Available as fiber, adhesive and other block or slab forms.

polycarbonate resins: High impact resistant thermoplastic polymers derived from linked dihydric polyester phenols.

polycondensation: See condensation.

polydispersity: Molecular weight distribution and nonuniformity.

polyester: A general description of resin formed by the reaction between a dibasic acid and a dihydroxy alcohol, both organic, or by the polymerization of a hydroxy carboxylic acid. Cure may be effected through vinyl polymerization using peroxide catalysts and promoters, or heat, to accelerate the reaction. (Thermoset or thermoplastic polyester)

polyether etherketone (PEEK): A linear aromatic high performance crystalline engineering thermoplastic.

polyetherimide (PEI): An armorphous thermoplastic polymer with good thermal properties.

polyethylene: A thermoplastic material composed solely of ethylene.

polyimide (PI): A highly heat resistant polymer produced by reacting an aromatic dianhydride with an aromatic diamine. May be either thermoplastic or thermoset.

polymer: A natural or synthetic, high-molecular-weight organic compound, whose structure can be represented by a repeated small unit, (the "mer"). Polymers may be formed by polymerization (addition polymer) or polycondensation (condensation polymer).

polymerization: A chemical reaction in which the molecules of a monomer are linked together to form large molecules whose molecular weight is a multiple of that of the original substance. When two or more monomers are involved, the process is called "copolymerization."

polymerize: The formation of a polymeric compound.

polymer matrix: Resin portion of a reinforced or filled resin system.

polymer reversion: The irreversible softening or liquifaction of a polymer as the result of hydrolysis.

polyol: Based on ethylene oxide/propylene oxide or propylene oxide. Produced as a reactant for the manufacture of foams.

polyphenylene sulfide (PPS): Engineering thermoplastic with outstanding chemical and heat resistance (450°F, or 230°C, continuous); excellent flame retardancy, low-temperature strength; inert to most chemicals over a wide temperature range.

polypropylene (PP): Thermoplastic resin with outstanding resistance to flex and stress cracking, good chemical resistance, electrical properties, impact strength and good thermal stability. Very low density.

polysiloxanes: Polymers synthesized by the polycondensation of silanols and containing Si-O linkage.

polystyrene: A water-white, low-cost, easy-to-process, rigid, brittle material; low moisture absorption, low heat resistance, poor outdoor stability; often modified to improve heat or impact resistance. Produced by polymerization of vinyl benzene.

polysulfides: Rubber type polymers containing sulfur and carbon linkages.

polysulfone: High heat-deflection temperature resin of melt-processible thermoplastics.

polytetrafluoroethylene (PTFE) resins: Chemically inert member of the fluorocarbon family of plastics. PTFE is made by the polymerization of tetrafluoroethylene, has extreme inertness to chemicals, very high thermal stability, low coefficient of friction and resists adhesion to most materials.

polyurethane resin: A thermosetting resin produced by reacting diisocyanate with organic compounds containing two or more active hydrogens to form polymers with free isocyanate groups. Applications include flexible molds, coatings and abrasion resistant castings. They exhibit good electrical properties and chemical resistance. Thermoplastic as well as thermoset polyurethanes (also urethanes) can be made into films, solid or foamed moldings, that are rigid or flexible.

polyvinyl acetate: The homopolymer or copolymer of vinyl acetate with more flexible monomers.

polyvinyl alcohol (PVA): A water soluble thermoplastic material composed of polymers of the hypothetical vinyl alcohol. The product is normally granular and is obtained by hydrolysis of polyvinyl esters, usually by the complete hydrolysis of polyvinyl acetate. It is used mainly for adhesives and coatings as well as fibers.

polyvinyl butyral: A colorless, flexible, tough thermoplastic solid material used primarily in interlayers for laminated safety glass. It is derived from a polyvinyl ester.

polyvinyl chloride (PVC): A colorless thermoplastic polymer synthesized from vinyl chloride monomer. Used to make fiber, insulation, fillers, outdoor components. Good water and chemical resistance. Compounded with plasticizers, it yields a flexible plastisol material that is superior to rubber for aging properties.

porosity: The percentage or ratio of the volume of air (void) within the boundaries of a solid material, to the total volume (trapped pockets of air, gas, or vacuum).

porosity, internal: The presence of numerous pits or pin holes beneath the surface; usually observable only in a cut cross section.

porosity, surface: The presence of numerous visible pits or pin holes at or near the surface.

porous molds: Molds that are usually used for vacuum or thermo-forming and are made up of bonded or fused aggregates including powdered metal, coarse pellets, etc. The resulting mass (usually cast) contains numerous open interstices of regular or irregular size through which either air or liquids may pass through the mass of the mold.

positive mold: Compression mold that will trap all molding material and prevent the escape of material being molded during the molding cycle.

post-cure: A "second stage" cure that follows the initial cure. Usually and without pressure in a mold, the post cure can occur "free standing" in an oven and ultimate properties are attained after exposure of the cured resin system to higher temperatures.

postforming: The operation used in the reinforced plastics industry to describe heating and reshaping of a partially or fully cured laminate. The formed or reformed laminate retains the contours and shape of the mold over which it has been formed.

pot life: Length of time, usually measured at room temperature ambient, before a catalyzed resin system has polymerized (hardened) to an unworkable state. The working or usable life before the resin system viscosity reaches the point of gelation.

potting: Similar to encapsulation, except the mold is a container that becomes part of the article that was cast in place inside. An encapsulated article is formed in a mold that is removed after casting.

potting compound: A material, usually an organic polymer, that is used for the encapsulation of components, wires and other articles.

powder: Particles of solid, small size matter within the range of 0.1 to 1,000 micrometers.

powder molding: Another description for slush or rotational molding where thermoset or thermoplastic powder material coats the inside of a mold to form a part.

power factor: The cosine of the angle between voltage applied and the current resulting.

precompaction: Application of pressure and/or vacuum, with or without heat, prior to curing, which removes trapped air, reactants, water vapor and/or volatiles from between plies of a layup. Generally used on thicker parts in order to reduce porosity of the final cured laminate. (Also compaction.)

precure: A partial or full setting of a resin system or adhesive in a joint before pressure is applied or the clamping operation is complete.

precursor: Organic fiber from which carbon fibers are made by pyrolysis using rayon, PAN or pitch fibers.

prefit: To check the fit of mating detail parts in an assembly prior to attachment by adhesive bonding or other means. Adhesive or mechanically fastened structures are sometimes prefitted to establish shimming requirements.

preform: A preshaped (contour and thickness) fibrous reinforcement of mat, cloth or tape formed to desired shape on a mandrel, mock-up or mold prior to being placed in the mold form for resin injection and subsequent cure. Reinforcement can be chopped fibers, knitted fiber, woven, continuous or other forms of reinforcement.

preform binder: A resin (thermoplastic or thermoset) applied to the chopped strands or fiber of a preform, usually during its formation, so that the preform will retain its shape and can be handled prior to insertion into the forming mold.

pregel: A layer of cured resin on part of the surface of a reinforced plastic part that does not relate to the "gel coat" and was unintentionally formed. (Not relating to "gel coat.")

preheating: Raising the temperature of a material(s) above the ambient temperature in order to reduce the thermal shock and to reduce the molding or dwell time during subsequent elevated-temperature molding or processing.

preimpregnation: The mixing of resin and reinforcing material before molding takes place.

premix: Molding compound prepared prior to and apart from the molding operations and containing all components required for molding; resin, reinforcement, fillers, catalysts, release agents, and other ingredients.

premold: Layup or molding and partial cure of a laminated or chopped-fiber part to allow handling, assembly and final cure with other parts.

preplastication: A technique of premelting and heating molding powders in a separate chamber, then transferring the melt to an injection cavity.

preplasticizer: Device used for preplastication.

preply: A composite material lamina with the composite reinforcement material of two or more plies prelaid in required position, ready for molding.

prepolymer: A chemical intermediate of rather low molecular weight is between that of the monomer or monomers and the final polymer or resin.

prepreg, preimpregnated: Ready-to-mold, continually reinforced material in sheet form which may be cloth, mat, or paper impregnated with resin and stored for use at a particular temperature. The resin is partially cured to a B-stage and supplied to the fabricator who lays up the finished shape and completes the cure with heat and pressure—either ready-to-mold material in sheet form or ready-to-wind material in roving form.

preproduction test: A test or tests conducted by a manufacturer or materials supplier to determine conformity to established standards.

presintering: Heating of a compact (powder metallurgy) to a lower temperature than the final sintering temperatures, to remove lubricant, bind and improve handling before sintering.

press: A device, usually mechanical or hydraulic that applies pressure on a material, part or assembly.

press polish: To form a high sheen or gloss finish on thermoplastic sheet stock by contact under heat and pressure in a press with highly polished platen plates. Also improves mechanical and other properties.

pressure: A Force measured per unit area. *Gauge pressure* is measured with respect to atmospheric pressure and absolute pressure is measured with respect to zero.

pressure bag molding: A composite molding technique where a flexible bag is placed over a contact layup on a mold, sealed, and pressure applied by compressed air, or steam, is trapped outside the bag pushing it against the part material while it cures.

pressure forming: Thermoforming shaping process wherein pressure pushes the sheet being formed against the mold instead of using vacuum to remove air beneath the sheet and cause it to go flat against the mold.

pressure intensifier: A layer or shape of flexible material (usually a high-temperature acrylic or silicone rubber) placed under or over a vacuum bag and used to ensure the application of increased pressure. Lo-

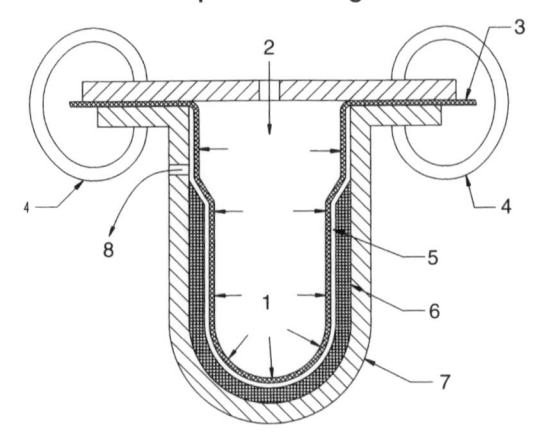

1—air pressure against bag
2—air pressure
3—pressure bag
4—clamp
5—bleeder, breather, ect. beneath pressure bag
6—mold laminate being formed
7—mold master laminate or master mold
8—vent to the atmosphere sometimes thru a vacuum pump

Pressure bag molding

cations include areas such as a radius, inside corner or recess in a layup being cured.

pressure pads: Reinforcements of hardened steel or other materials, distributed around dead areas on the faces of a mold to help absorb the final pressure of closing without collapsing. Elastomeric pads are sometimes used taking advantage of the increased coefficient of thermal expansion when heated with the material being formed.

pressure-sensitive adhesive: A permanently tacky material that will adhere instantaneously to most adherend and solid surfaces with the application of very light pressure at room temperature.

prestite: A vacuum bagging clay-like sealant strip used to seal bags to mold surfaces.

primary plasticizer: A material that can be used as the sole plasticizer if its compatible with the resin system or polymer.

primary structure: Usually on a flight vehicle, the major components and assemblies critical to flight safety.

primer: An initial coating applied to a surface, prior to the application of an adhesive or other coating that protects that surface and improves the performance and bond of subsequent coatings.

principal direction: The orientation of specific coordinate axes when stress and strain components are maximum and minimum for normal components, and zero for shear.

printed wiring or circuit board: A description of completely processed-printed wiring and printed circuit board configurations. These include single-sided, double-sided and multilayer boards with rigid, flexible, and semirigid substrate. Electrical connections and sometimes components are formed by additive wire or metal patterns or reduction by metal etching electrical pattern connections including through holes to the lower layers and opposite side of the board.

profile: Contour of the surface of a substrate as viewed from the crossection or peripheral shape edge.

profile die: Die or shaping cavity for the continuous forming shapes, (not sheets or tubes).

promoter: A chemical, that increases and accelerates activity of a catalyst or initiator.

prototype: A model used to demonstrate the design, form, performance and other aspects.

prototype mold: A simple mold usually made from a light casting metal alloy or mass cast epoxy, laminate, plaster or either cost effective material as a temporary means of forming shapes.

pseudoplastic: A fluid whose viscosity or consistency decreases instantaneously with an increase in shear rate. Relatively high resistance to stirring will decrease as the rate of stirring is increased.

puckers: An area or areas on prepreg materials where a local blister has caused separation from the release film paper.

pulp: Cellulose obtained from wood or other vegetable matter.

pulp molding: A molding process where resin-impregnated pulp materials are preformed by application of a vacuum and subsequently oven cured or molded.

pulform (pulpress): Multistage pultrusion process where press forming is performed following the pultrusion stage.

pultrusion: Continuous composite forming process for shaping rods, tubes, and structural shapes of constant cross section. Pre-impregnated reinforcement is pulled through a shaping and curing die to the desired cross section.

puncture: A break in the skin that may or may not extend through inner and outer skin.

pyrolysis: A thermal process by which organic precursor fiber materials, such as rayon, polyacrylonitrile (PAN), and pitch, are chemically changed into carbon fiber by the action of heat (800 to 2800°C, 1470 to 5070°F) in an inert atmosphere. Graphitization processing temperatures of 1900 to 3000°C (3450 to 5430°F) result in higher modulus carbon or graphite fibers.

Q

qualitative analysis: An identification of materials by a subdivision of chemistry.

qualification test: Testing conducted by an organization, procuring activity, or its agent, resulting in conformance of materials, system, product acceptance to the requirements of a specification.

quality assurance (QA): Procedures confirming conformance of material or product specification qualification.

quality conformance (Control): Qualification performed on a regularly scheduled basis in order to demonstrate the continued ability of a material, process, workmanship and product to meet all of the quality requirements specified.

quantitive analysis: The percentage composition of mixtures or the components of a pure compound determined by a subdivision of chemistry.

quartz fibers: Fibers produced from high quality and purity natural quartz crystals (99.5% SiO_2).

quasi isotropic laminate: A laminate that has plies oriented and usually balanced in all directions (0 to −45 to 90 to +45°) (0 to +60°). Physical and mechanical properties in the laminate plane are almost "quasi isotropic."

quench (thermoplastics): Shock cooling process for thermoplastic materials that are in the molten state.

quench bath: The cooling medium used to quench molten thermoplastic materials to the solid state as in the extension of film through a water bath.

R

radiation: Electromagnetic energy transmission or emission.

radiation curing: Use of electromagnetic (electrons, protons, etc.) energy to cure polymers.

radiation damage: The embrittlement, molecular bond severance or degradation of properties due to ionizing or penetrating radiation.

radio frequency: Electromagnetic radiation detectible around 10 to 100 GHz.

radio frequency interference (RFI): Interference caused by radio frequencies.

radio frequency preheating (RFP): A heating method to heat materials to be molded, using R.F. energy prior to the molding operation.

radio frequency welding (RFW): Method of heating, connecting or welding thermoplastics using radio frequency energy (high frequency welding).

radiometry: Measurement of infrared, ultraviolet and visible radiation in the optical spectrum.

ram: A machine or forming press member that exerts pressure by entering the cavity block. Molding material or compound is forced (rammed) into mold cavity or cavities.

ram force: Total load equal to product of line pressure and the cross-sectional ram area as applied by the ram. It is normally expressed in tons.

ramping: The programmed process for gradual increase or decrease of pressure or temperature of an autoclave, oven, tool or combination thereof when molding composite or plastic parts.

ram travel: For transfer or injection molding, the distance the injection ram moves heated materials when totally filling the mold.

random copolymer: Usually results from the copolymerization of two monomers and the molecules of each monomer are randomly arranged in the polymer backbone.

random effects: Unavoidable inherent process changes of quality characteristics.

randomness: A situation in which any individual event has the same mathematical probability of occurring as does all of the other events within the same set of events.

random pattern: A winding in which the filaments do not lie in an even pattern. A winding with no fixed pattern.

random sample: The selection of individual items or units where each possible individual unit has an equal chance of being selected.

ranking: Presenting items, values, properties of materials in a precise order such as a laminate by strength, stiffness or others.

rayon: A generic term for regenerated cellulose fibers. The chemical structure of rayon fibers are similar to natural cellulose fibers.

reaction injection molding (RIM): An injection molding process using activated thermosetting resins and lower pressures than for thermoplastics with (usually) a high-pressure impingement type mixing head.

reactivity ratios: In a copolymerization reaction, the ratio of the rate of reaction of a monomer with itself as compared to the rate of reaction with its comonomer.

rebond: The bond or termination made at, adjacent to, or on top of, the location of the original bond.

receiving inspection: Inspection by recipient upon receipt of material or manufactured products.

reciprocating screw: An injection machine device where there is injection of material or melt into a die or mold by direct extrusion into the mold, or by reciprocating a screw as an injection plunger, or by a combination of the two. The screw can serve as an injection plunger.

recycle: Reprocess ground material from original parts, flash and trimmings which, after mixing with a certain amount of virgin material, is remolded.

reduced inspection: Inspection, less severe than the normal inspection. Used when the results of normal

reference dimension

inspection of lots/batches indicate that the quality of the product is better than that specified.

reference dimension: A dimension with no a tolerances, used only for informational purposes that do not govern inspection or other manufacturing operations.

reflection: The amount of reflector or returned radiant energy from a surface.

reflow soldering: A joining method of mating surfaces that have been previously coated with solder or tinned, heated and cooled after solder reflow.

refractory: A heat-resistant material or alloy of very high melting point.

reinforced molding compound: Compound of ready-to-use materials supplied by raw material suppliers, as distinguished from premix. *See also* Premix.

reinforced plastic: A plastic with reinforcements imbedded in the composition and strength properties greatly superior to those of the base resin. Formed, molded, tape wrapped, filament wound or shaped composite parts consisting of resins to which reinforcing fibers, mats, fabrics, have been added.

reinforced reaction injection molding (RRIM): A reaction injection molding with a reinforcement added. *See also* reaction injection molding.

reinforcement: A strong material added to and bonded into a plastic matrix to improve its strength, stiffness, and impact resistance. Reinforcements are usually inert woven and nonwoven fibers of boron, carbon, ceramic, cotton, flock, graphite, jute, sisal, paper, or other synthetic materials. Reinforcement is not a filler and provides improved tensile/flexural strength.

rejection number: In sampling inspection by attributes, the lowest number of defects or defective items found in the sample that requires the rejection of the lot.

relaxation time: The time required for a stress under a sustained constant strain to diminish by a stated fraction of its initial value.

relaxed stress: The initial stress minus the remaining stress at a given time during a stress-relaxation test.

release agent: A material that is applied in a thin or thick film to the surface of a mold or tool to keep the resin system from bonding to the surface. Also called parting agent or mold release agent.

release film: A sheet of material or impermeable layer of film placed adjacent to a mold surface to minimize adhesion between the mold and the curing resin system.

release paper: A sheet, serving as a protectant, carrier, or separator, for an adhesive film, prepreg or mass, which is easily removed from the film, prepreg or mass prior to use.

relevant failure: Failure that is to be included interpreting test results or in calculating the value of a reliability characteristic.

reliability (international quality term): The ability of an item to perform a required function under stated conditions for a stated period of time.

reliability compliance test: An experiment carried out to show whether or not the value of a reliability characteristic of an item complies with its stated reliability requirements.

reliability data: Data on characteristics permitting quantitative evaluation of reliability.

reliability determination test: An experiment carried out to determine the value of a reliability characteristic of an item.

repair(ing): The act of restoring the functional capability of a defective article in a manner that precludes compliance of the article with applicable drawings or specifications.

repeating index: A laminate code to represent the number of repeating sublaminates.

residual gas analysis (RGA): The study of residual gases in vacuum systems using mass spectometry.

residual strain: The strain associated with residual stress.

residual stress: The stress existing in a body at rest, in equilibrium, at uniform temperature, and not subjected to external forces. Often caused by the forming and curing process.

residue: A visual or measurable form of process-related contamination.

resilience: Ability to quickly regain an original shape after being strained or distorted.

resin: Solid, semisolid, or pseudosolid organic material, generally a polymer, which has an indefinite (often high) molecular weight, exhibits a tendency to flow when subjected to stress, usually has a softening or melting range, and used to bind the reinforcement material in composites, form any adhesive, casting or potting material base.

resin applicator: A device used to deposit a liquid resin system. In filament winding, used to apply the resin onto the reinforcement band.

resin content: The amount of matrix present in a composite either by percent weight or percent volume.

resin pocket: A visual accumulation of excess resin in a small, localized section visible in angles, recesses and on cut edges of molded surfaces.

resin-rich area: Location or space such as radii, steps, etc., which is filled with resin and lacking fiber reinforcment material.

resin ridge: A sharp buildup on a part surface consisting of only resin.

1–female mold
2–resin pocket
3–laminate lay-up

Resin pocket

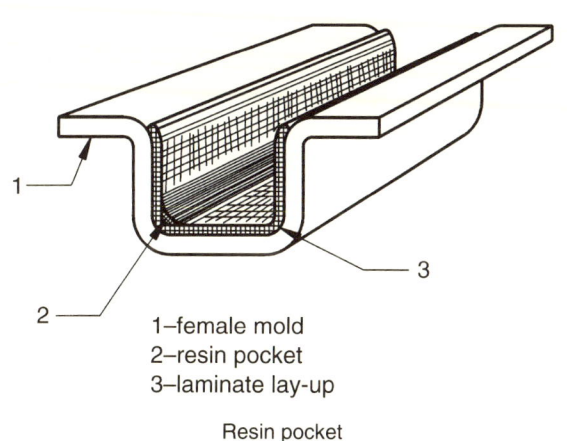

resin-starved area: Area deficient in resin, usually identified by voids, dry spots, or loose fibers.
resin system: A mixture of resin and ingredients such as hardener, curing agent, catalyst, accelerator, initiator, diluents, fillers, and so forth, required for the intended processing and final product.
resin transfer molding (RTM): A closed mold, low-pressure injection molding process whereby catalyzed resin is transferred or injected into a closed mold in which the fibrous reinforcement form has been placed. Reinforcement forms include preshaped woven and nonwoven glass, graphite or aramid preforms. Resin systems include all thermoset materials such as polyester, vinyl ester, BMI, epoxy and others.
resinoid: Any class of thermosetting synthetic resin system, either in the initial temporarily fusible state or in the final infusible state.
resist: A coating material that is used to mask or protect selected areas of a surface from the action of a chemical or other processing reaction.
resistance welding: Joining process by fusion welding using pressure and heat generated by resistance heating.
resistivity: The resistance to passage of electric current along the surface or through the bulk of a material. The unit of volume resistivity is the ohm-cm, or surface resistivity, the ohm.
resite: An alternative term, a fully cured thermoset C-staged resin system.
resitol: The partially cured or B-stage in thermoset resin systems.
resol: A resin system which is in its uncured and ready-to-form A-stage state.
retarder: An additive that slows down a chemical reaction or physical change.
reticulated foam: An open structure foam where the cells are shaped by a fine thread network.

reverse helical winding: In filament winding and in contrast to biaxial, compact, or sequential winding, as the fiber delivery arm traverses one circuit, a continuous helix is laid down, reversing direction at the polar ends. Fibers cross each other at definite equators, the number depending on the helix, with three as the minimum region to cross over.
reverse impact test: A sheet material test where one side is struck by a falling object or pendulum. The reverse side is inspected for damage.
reversible change: Change in which a characteristic value is unambiguously correlated with an influencing quantity.
reversion: When a resin system (polymerized material) partially or completely degenerates through significant changes in physical and mechanical properties to a lower polymeric state or to the original monomer.
rework: The act of reprocessing a material or item, through the use of original or alternate equivalent processing
R-glass: European equivalent of S-glass.
rheology: The science for deformation and flow of matter, the study of the flow of materials, particularly plastic flow of solids and the flow of non-Newtonian liquids on a macroscopic and microscopic level.
rib: A stiffening or reinforcing member of a fabricated or molded part.
ribbon: A cross section shape (i.e., fiber) having a rectangular cross section, where the width-to-thickness ratio is at least 4:1.
ribbonization: The degree of flattening of a sized roving, expressed as the ribbon width to thickness ratio.
rigid: Stiff with no flexibility.
rigid plastics: A plastic that has a modulus of elasticity either in flexure or in tension greater than 690 mpa (100 Ksi) at 23°C (73°) and 50% relative humidity.
rise time: For polyurethane foam molding, the time from pouring of the urethane mix to the completion of foaming.
risk: A combined effect of the probability of occurrence and magnitude of an undesirable event.
risk management: The process whereby decisions are made to accept, eliminate or mitigate a known risk or hazard.
rockwell hardness: A testing method to measure resistance to indentation. A diamond cone and steel ball, under pressure, are used to pierce the test specimen.
roller: A serrated tool used in hand lamination processing to compact a wet laminate and release trapped air or volitiles.

roll mill: A machine with two rolls placed in close relationship. Used to admix a plastic material with other substances by having the rolls turn at different speeds to produce a material shearing action.

room temperature: 60° to 100°F (usually 77°F or 25°C).

room temperature-curing adhesives: Adhesives that set to handling strength in about an hour at 68 to 86°F (20 to 30°C) and reach full strength without heating.

room temperature vulcanizing (RTV): Chemical reaction, curing, or vulcanization at room temperature (usually applies to silicones and other rubbers).

rosin: A hard, natural resin, that is extracted from the sap of pine trees and subsequently refined.

rotational molding (rotomolding): Method to make hollow articles from thermoplastic materials (powder) or thermoset (powder or liquid). Material is charged into hollow mold capable of being rotated in one or more planes. The hot mold fuses or cures the material after the rotation has caused it to cover all mold surfaces. The mold is then chilled and the product removed.

roving: A number of tows, twisted yarns, untwisted strands, or ends collected into continuous parallel bundles usually used in the filament winding process.

roving ball: A winding machine supply package consisting of a continuous number of ends or strands wound to a given outside diameter onto a round mandrel.

roving cloth: A course textile fabric, woven from rovings.

roving integrity: Degree of bonding between roving strands.

rubber: An elastomer capable of high elastic deformation and elongation. Specifically, heavy or natural rubber, the standards of comparison for elastomers.

rule of mixture: The relationship equation that calculates certain physical and mechanical properties of a given laminate based upon the properties of the constituents (resin and fiber) and their volume fractions in that laminate.

run: A consecutive number of points that consistently increase or decrease, in a series of observations.

runner: A main feed channel, usually circular, or material formed in a channel of an injection or transfer mold connecting the sprue with the gate to the cavity.

runner system: Term applied to all the material in the form of sections of the channel system that includes the runners, sprues, and gates which lead material from the nozzle or pot, to the mold cavity.

rupture: A break, cleavage, or burst as a result of physical stress.

S

safety hardener: A curing agent which causes only a minimum of toxic effects on the human body, either by contact with the skin or as concentrated vapor in the air.

safety management: The application of organization and management principles in order to assure with high confidence the timely realization of the goal of optimum safety.

sagging: Description of a wet coating or paint film to flow down and become thicker in some areas.

sample: A group of items or individuals, taken from a larger collection or population, that provides information needed for assessing a characteristic (or characteristics) or the population, or which serves as a basis for action on the population or the process that produced it.

sampling inspection: The inspection of a limited number of items or of a limited quantity of material, taken at random from the lot or batch according to a prescribed sampling plan.

sampling plan: A statement of the sample sizes and decision criteria (e.g., acceptance or rejection numbers for inspection by attributes) applicable to a particular lot/batch in accordance with a particular sampling scheme.

sampling scheme: An overall system containing a range of sampling plans and procedures, based on the mathematical theory of probability and statistics whereby the results of inspecting one or more samples is used to determine the acceptance or rejection of a lot/batch, or to assess its quality.

sampling size: The number of specimens in the sample.

sandwich construction: A structure (composite) composed of a lightweight core material (honeycomb, foamed plastic, etc.) to which two thin, dense, high strength faces or skins are adhered.

sandwich heating: A heating method for processing a thermoplastic sheet by heating both sides.

saran plastics: Group of plastics whose resins are derived from the polymerization of vinylidene chloride or the copolymerization of vinylidene chloride and other unsaturated compounds.

satin weave: A warp-faced weave in which the binding places are arranged with a view to produce a smooth cloth surface free from twill, with an irregular interlacing shift that can conform to compound cured surfaces.

saturated compound: Organic compounds that cannot add on elements or compounds.

S-basis: The S-basis is a minimum property allowable.

scale: A surface condition where particles can be observed on the surface of a composite or plastic.

scarf joint: A overlapped and bonded joint in which similar segments of adherents are cut away.

scratch: Shallow surface mark usually caused by improper handling or storage.

screening inspection: A 100% inspection carried out for the purpose of removing defective items from a lot that has been rejected.

scrim: A low-cost reinforcing open-mesh fabric made from continuous-filament yarn and used to support B-staged adhesives.

sealant: Material applied to a surface or joint as paste or liquid which hardens or cures forming a seal against gas or liquid entry.

sealant tape: Tape (elastomeric) that seals the vacuum bag to the model, mold or tool and also seals bag to bag surface.

secant modulus: The ratio of total stress to corresponding strain at any specific point on the stress-strain curve, expressed in force per unit area.

secondary bonding: A process of joining two or more already formed, cured or bonded nonmetallic parts (and possibly other metal fittings/parts) using adhesive bonding.

secondary failure: Failure of an item caused either directly or indirectly by the failure of another item.

secondary structure: Not critical to flight safety of an aircraft or other similar structures.

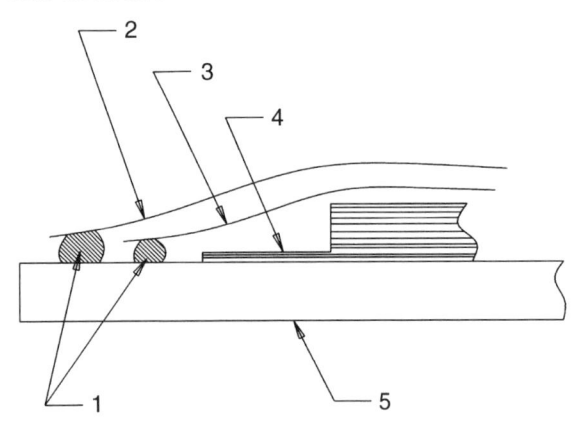

1–bag sealant tape
2–vacuum bag #2
3–vacuum bag #1
4–mold or tool laminate
5–master model

Sealant tape double vacuum bag

self-destruct: The temperature and moisture level required to cause structural failure without any externally applied stress.

self-extinguishing resin: A resin formulation that only burns in the presence of a flame. It will extinguish itself within a specified time after removal of the flame.

self-skinning foam: Foam that produces a tough outer surface over the foam core upon formation.

selvage: The edge of a woven fabric running parallel to the warp that prevents raveling.

semicrystalline: Plastic materials exhibiting localized crystallinity.

semipositive mold: One where a small amount of extra material escapes when it is closed resulting in close tolerance parts.

semirigid plastic: *See* plastic.

separate application adhesive: One that has two components with each applied to the opposite adherends.

separator: Permeable layer that acts as a release film such as porous teflon-coated fiberglass.

sequential sampling: A type of sampling that requires the taking of successive items, or sometimes successive groups of items, but without fixing their number in advance, the decision to accept or reject the lot/batch being taken as soon as the results permit, according to the sampling plan.

sequential winding: *See* biaxial winding.

serving: Wrapping a yarn such as rayon around a roving or yarn for protection.

set: (1) Conversion into a hardened or fixed state using chemical or physical action, such as polymerization, condensation, oxidation, gelation, vulcanization, evaporation, or hydration of volatiles. (2) Creep or deformation usually expressed as a percentage of original dimension (irrecoverable).

set at break: A measurement of elongation on a reassembled tensile specimen ten minutes after rupture.

setting time: The time period when an adhesive, potting compound, encapsulant or other thermoset material becomes set and ready to cure.

set up: A condition of a thermoset polymer, to harden, to cure.

set up time: The period of time during which a molded or extruded product is subjected to heat and/or pressure to set the resin or adhesive.

S-Glass: A glass of magnesium, aluminia, silicate composition (structural) that is designed to provide very high tensile strength filaments (S-2 is commercial grade).

shear: A stress action that results from applied forces and will cause or tend to cause two contiguous parts of a body to slide parallel and relative to each other's plane of contact.

shear coupling: Unique with anistropic materials, this induced shear strain is a result of normal stress.

shear edge: In a mold it is the cutoff edge.

shear modulus: Ratio of shearing stress to shearing strain within a materials' proportional limit.

shear strain: Also called angular strain, is the tangent of the angular change, caused by a force between two lines originally perpendicular to each other through a point in a body.

shear strength: Maximum shear stress a material is capable of sustaining.

shear stress: Component of stress tangent to the plane on which the forces act.

sheet: Thermoset or thermoplastic flat sections where the length is considerably greater than the width.

sheet molding compound (SMC): A thermosetting resin system, fibrous reinforcement and pigments, fillers, and other additives compounded and processed into a sheet form that can be molded net, by comparison, into compound and complex curves and shapes.

sheet train: Assembly or tooling necessary to produce flat sheet. Includes extruder, die, polish rolls, conveyor, draw rolls, cutter and stacker.

shelf life: The period of time a material or product can be stored in a specified environment and still meet all requirements and remain acceptable for its original use.

shell tooling: A mold or bonding fixture having a contoured shell surface supported by a backup which assures dimensional stability.

shelling: A term applied to loops of roving falling to the base of a roving ball as the roving is paved out.

shoe: A filament gathering device for glass fiber strand forming.

shore hardness: The resistance of a material to indentation by measuring using a spring loaded indenting device. The higher number the greater the resistance.

short-beam shear strength: Interlaminar shear strength of a parallel fiber reinforced material using a three-point flexural specimen loading.

short fiber: A fiber length which is less than the critical length, and hence does not achieve full reinforcing efficiency.

shortness: Qualitative term to describe an adhesive that does not string or form filaments or threads during application.

short shot: Injection molding of insufficient material to fill the mold.

shot: The total amount of material or yield from one complete molding cycle.

shot capacity: Maximum weight of material an injection machine can provide on forward motion of the ram, screw, or plunger.

shrinkage: The dimensional difference between the mold geometry and the cured component geometry 24 hours after removal from the mold.

shrink fixture: Device upon which molded parts are placed so that they maintain shape during cooling.

silane: Adhesion prometer or coupling agent used on the material or part surface, reinforcement or with the resin system.

silica: High purity glass or sand (silicon dioxide) used as a filler.

silicone: Derived from silica and used as a release agent and general lubricant.

silicone bag: Permanent reusable vacuum bag made of silicone, elastomer sheet and edged sealed with an inner locking seal.

silicon carbide: Hexagonal crystals (abrasive) of silicon carbide.

silicone carbide fiber: A high strength and modulus reinforcing fiber made by a vapor-chemical deposition process.

silicones: Resinous organosiloxane polymer materials based on resins where the main polymer chain consists of alternation in silicon and oxygen atoms and with carbon-containing side groups, available in different molecular weights including liquids, solid resins and elastomer.

single-circuit winding: Filament winding in which the filament path makes a complete traverse, and the following traverse lies immediately adjacent to the previous one.

single load: Stress applied to a body in only one direction.

single point milling: Conventional method of shaping or dressing a die using a single rotary cutter on a machine tool.

single sampling: A type of sampling that consists of taking only one sample from a lot/batch.

single-sided board: Printed wiring or substrate board with conductive pattern on only one side.

sink mark: Shallow depression on the surface of a molded part due to collapsing of the surface following local internal shrinkage.

sintering: In powder metallurgy the bonding of adjacent surfaces of particles in a mass by heating with or without prior compacting.

size: A numerical description of the dimensions describing an article. Surface treatment (finish) applied to reinforcement filaments before or during forming operations. Sizes provide good handling and adhesion characteristics.

sizing: Compound that binds together and stiffens yarn, providing resistance to abrasion during weaving; normally removed and replaced with finish before matrix application. The material used for this purpose, is called size.

sizing content: The percent of the total strand weight made up by the sizing, as determined by burning off the organic sizing and measuring weight.

skein: A continuous strand, yarn or roving wound up to some measured length and usually used to measure various physical properties of the material.

skin: A layer of relatively dense material used in a sandwich construction on the surface of the core or cellular plastic.

skirt: The material added to the end(s) of a filament wound bottle to provide attachment points for supports. Extension of the cylindrical portion of a rocket motor case from the equator, used for interstage connections, usually wound as an integral part of the case.

slave skin: A plate located to the top of a component during cure to achieve a uniform pressure distribution. Also known as caul plate.

slave skin/caul plate: A plate located on the top of a component during cure to achieve uniform pressure distribution.

slip: Creamy consistency thinned ceramic powder or clay. Also sometimes referred to as "glide" or movement of one item along a plane.

slip angle: The angle at which tensioned fiber will slide off a filament wound dome. The value is dependent on the fiber and resin system used.

slip forming: A sheet-forming technique.

slippage: During the bonding process the movement of adherends with respect to each other.

slipper skin: Composite skin used in conjunction with metallic tooling to account for mismatch of thermal expansion between the tool and component during cure.

sliver: A continuous strand (without twist) composed of staple or continuous filament fibers.

sluffing: In the pultrusion process a condition wherein scales peel off or become loose.

slurry: An insoluble watery mixture of matter, such as plaster.

slurry preforming: Method for preparing fibrous preforms by wet processing collection of fibers on a screen form and evacuation of slurry-water through the screen.

slush molding: Thermoplastics, casting method where the resin in liquid form is poured into a hot mold and a viscous skin forms. Excess is drained off, the mold is cooled, and the molded part is stripped out.

snap back forming: A thermoplastic sheet-forming technique.

S-N curve: Stress per number of cycles to failure. *See also* stress-strain.

softening range: Temperature range in which a plastic transforms from a rigid solid to a soft state.

soft flow: Behavior of a material flowing freely under conventional molding conditions and fills all the interstices of a deep mold with a considerable distance of flow.

solid laminate: A resin-impregnated structurally reinforced cured composite laminate in a solid state, containing no sandwiching layers.

solids content: The percentage by weight of nonvolatile material.

solvation: Process of swelling of a resin or plastic caused by interaction between a resin and a solvent or plasticizer.

solvent: A substance (usually liquid) in a solution used to dissolve another substance or for cleaning materials during manufacturing operations.

solvent activated adhesive: Dry adhesive film that becomes tacky just prior to use by applying a solvent.

solvent adhesive: An adhesive having a volatile organic liquid as a vehicle.

solvent bonding: Connecting or joining thermoplastic surfaces by application of an appropriate solvent.

solvent cleaning: Removal of organic and inorganic soils using blended polar and nonpolar organic solvents.

solvent resistance: Ability of a plastic to withstand any type of solvent attack.

specification: The detailed description of the characteristics of a material, process or product including criteria used to determine conformity with the description.

specific gravity: The ratio of the weight of any volume of a substance to the weight of an equal volume of another substance measured as standard at a constant or stated temperature (density, mass per unit volume).

specific heat: Amount of heat required to raise the temperature of a substance unit of specified mass one unit of specified temperature (Cal/g/°C or Btu/lb./°F) 1 degree under specified conditions.

specific modulus: The modulus-to-density ratio.

specific properties: The material properties divided by material density.

specific strength: The strength-to-density ratio.

specific stress (fibers): Load divided by mass per unit length of the test specimen.

specimen: A representative single item, piece, part of sample or a measured quantity of material used to make a specific test.

SPI gel time: The interval of time between introduction of the catalyst to a thermosetting resin and the formation of a gel, as defined by the U.S. Society of the Plastics Industry.

spinneret: An extrusion die or metal plate containing many tiny holes, through which a melted plastic is forced forming fine fibers and filaments which are hardened by cooling in air, water, or chemical action.

spinning: Process of making fibers by forcing plastic melt through spinnerets.

spiral: When forming glass fiber this device is used to make the strand traverse back and forth across the forming tube.

spiral flow test: A measure of the moldability of SMC.

splay: Fanlike surface defect near the mold gate on a part.

splice: The joining of two ends of a yarn, tow or strand, by interlacing or adhesive.

spline: A tool or the process of preparing a surface to a desired contour by working a paste material with a flat-edge tool similar to screeding plaster or concrete. A method for engendering solid geometry from plane geometry.

split cavity: Mold cavity made in sections.

split-cavity blocks: Blocks that are used in a mold cavity for molding articles having undercuts.

split mold: A mold in which the cavity is formed of two or more components (splits).

split-ring mold: Mold in which a split-cavity block is assembled in a chase. Parts having undercuts are

1–molded part
2–(3)piece mold

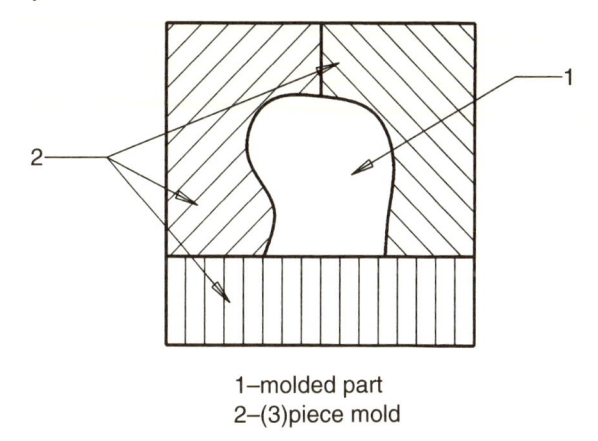

Split mold (3) parts

ejected from the mold and then separated from the split-cavity block assembly.

spool: A cylindrical shape on which various materials are wound. Also, a term used to identify a roving ball is the preferred term.

spray: Complete set of molded parts from a multi-impression injection mold, together with the flash, injection sprue and other molded material. Also method of applying coatings.

sprayed metal molds: Molds made by spraying vaporized or molten metal onto a master model or mandrel until a mold surface of predetermined thickness is achieved. The metal shell is backed up with a laminate, plaster, cement, casting resin, or other suitable material. Can produce near "Class A" surface finishes and can be used to mold thermoplastic and thermoset materials.

spray-up: Method in which a modified spray gun is used as an applicator tool to. Apply chopped fibrous reinforcement and resin system as in reinforced plastic processing.

spread: A quantity of adhesive applied along a joint area (or areas) of an adherend, usually expressed in pounds of adhesive per thousand square feet of joint area.

springback: The movement in or out of the edge of a molded part following part removal from the mold.

spring constant: In a prescribed test procedure, the number of pounds required to compress a specimen one inch.

sprue: The hole pattern through which thermoset molding compounds are injected directly into the mold cavity. Also the molding compound shape resulting from removal of the part.

spun roving: Low-cost heavy glass-fiber strand of filaments that are continuous but doubled back on each other.

square symmetry: Refers to a reinforcement that has equal stiffness or strength properties in two orthogonal axes (such as a square weave fabric).

squash molding: A molding process accomplished by pressing together two or more mold parts. Excess resin is "squashed" out through the mold parting lines.

squeeze out: Excess adhesive pressed out at the bond line between adherends.

stability: Ability of a material to resist and maintain resistance to various elements and environments.

stabilization: A carbon fiber process used to make the fiber precursor infusible prior to carbonization.

stabilizer: Ingredient used to assist in maintaining the physical and chemical properties of the compounded materials.

stacking sequence: A laminate description in which ply orientations or individual laminae are defined as laid down (or stacked) into a laminate.

staged prepreg: Partially advanced cure of thermoset matrix material allowing room temperature storage prior to final cure.

staging: Heating premixed resin system to start curing and stopping the reaction before the gel point. Particularly used in the production of prepregs.

stamping: The rapid application of force to sheet feedstock material to produce components of shallow depth.

standard deviation: The root mean square of the individual deviation from the mean value as a measure of dispersion of data from the average.

standard laboratory atmosphere: One where there is a relative humidity of 50 ± 2% at a temperature of 23 ± 1 C (73.4 ± 1.8°F), where the Average room conditions are 40% relative humidity at a temperature of 77°F. Dry room conditions are 15% relative humidity at a temperature of 85°F. Moist room conditions are 75% relative humidity at a temperature of 77°F.

standby time: The period of time during which an item is needed to be in a condition to perform its required function but is not operated.

staple fibers: Short fibers of spinnable length manufactured directly or by cutting continuous filaments to relatively short lengths that can be spun into yarn.

starved area: Area in a reinforced composite part that did not have sufficient matrix resin system to completely wet the reinforcement.

starved joint: Adhesive joint which does not have the proper film thickness of adhesive due to insufficient adhesive spreading or to the application of excessive pressure during the bonding process.

static fatigue: Usually aging accelerated by stress causing a failure of a part under continued static load.

static modulus: The ratio of stress to strain in shear, compression, or tension under static conditions.

static stress: Stress in which the force is constant and slowly increasing with time, such as, test of failure without shock.

step cure: Cures that start at lower temperatures and incrementally increase up to the cure temperature.

stereospecific polymers: Specific or definite order of arrangement of molecules.

stiffened panel: Panel with reinforced skin members located in structural directions.

stiffness: The measure of modulus and relationship of load and deformation (modulus of elasticity).

stoichiometry: Control of levels, and exact amounts in a chemical mix necessary for a proper reaction.

stops: Blocks inserted between die halves that control thickness of a press-molded part.

storage life: The specified length of time prior to use for which materials or other items are inherently subject to deterioration and are deemed to remain fit for use under prescribed conditions.

strain (ϵ): Although strain has several definitions, which depend upon the system being considered, for small deformations, engineering strain is applicable and is the most common definition of strain. The quality called true strain is sometimes used in areas of plastic deformation and is the geometric measurement of deformation.
  (1) Engineering strain (ϵ)
      The ratio of the change in length, ΔL, of the sample to its original, L_0.
  (2) True strain (ϵ_1)
      The integral of the ratio of the incremental change in length to the instantaneous length of a plastically deformed sample; thus, the natural logarithm of the ratio of instantaneous length to original length of such a sample.

strain gauge: A device (extensometer) used to measure strain in a stressed material based on the change in electrical resistance of metal elements attached to the nonmetallic sample surface.

strain invariant: The scaled combination of strain components.

strain rate: In mechanical testing operations, the change in strain per unit time.

strain relaxation: *See* creep.

strand: A single (monofilament) or untwisted bundle or assembly of continuous filaments, tows, yarn, ends, including slivers, and used as a unit.

strand count: The number of strands in a plied yarn or roving, such as denier.

strand integrity: The degree of bond between the filaments in a strand as a result of the filament sizing adhesion.

stratified sampling: A type of sampling in which the population is divided into different subpopulations (called "strata") and sampling is then carried out so that specified proportions of the sample are drawn from the different strata.

strength: Maximum strain, stress and rupture that a material can sustain.

strength, dry: Strength of an adhesive joint determined immediately after drying under specified conditions or after conditioning in a standard laboratory atmosphere.

strength, wet: Strength of an adhesive joint determined immediately after removal from a liquid. Tests are performed under specified conditions of time, temperature, and pressure.

strength in compression: Maximum load sustained by a specimen divided by the original cross-section area of the specimen.

strength in tension: Ability to withstand deformation under a tensile load.

stress (σ): The study of the effects of stresses on an item and their distribution, taking into account the operational requirements of the item. Most commonly defined as "engineering stress," the ratio of the applied load P to the original cross-sectional area A_0.

stress concentration: The applied stress with increased ratio of a local stress over the average stress. At a macromechanical level, the concentration occurs at holes, voids, notches, ply termination points, joints and corners. At a micromechanical level the concentration occurs at the fiber/matric interface.

stress concentration factor: Ratio of the maximum stress in the region of stress concentration, (hole, void, point) to the stress in a similar strained area without stress concentration.

stress corrosion: The degradation of an area under stress in a corrosive environment, where the environment alone could not cause corrosion.

stress crack: Internal or external defect in a plastic caused by tensile stresses.

stress cycle: Repeatable sequence of stresses.

stress raisers: Local increases in stress caused by changes in contour or discontinuities in the structure.

stress relaxation: A decrease in stress (decay) under sustained constant strain (creep).

stress-strain: The stiffness (pounds per square inch or kilograms per square centimeter) at a given strain.

stress-strain curve (diagram): Plotting the applied stress using simultaneous readings of load and deformation, converted to stress and strain.

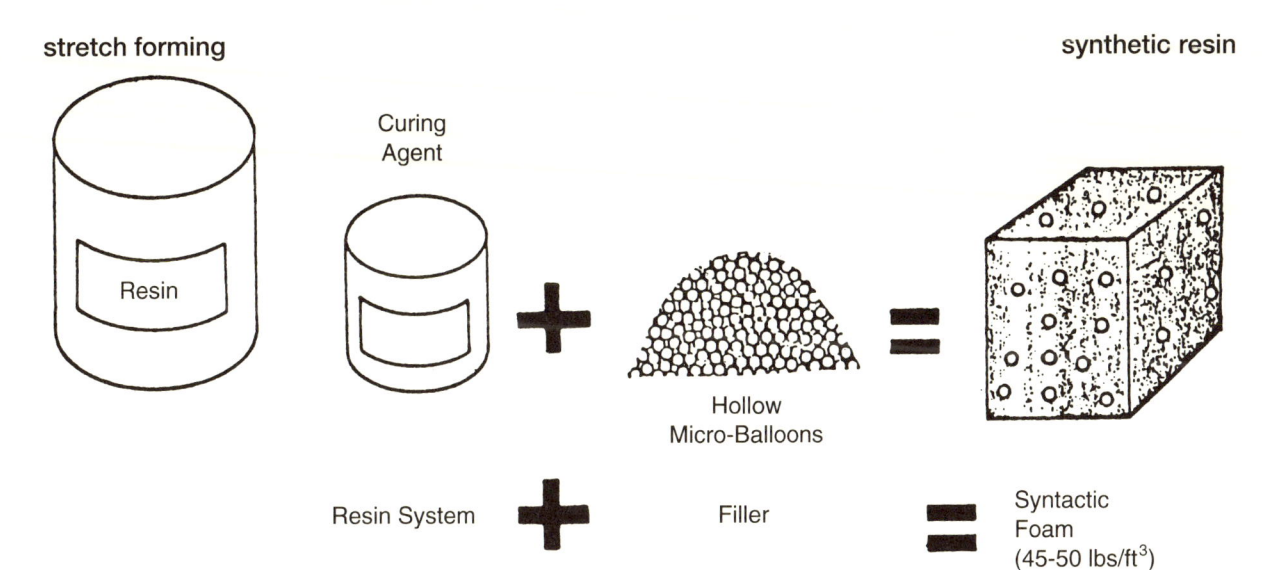

Syntactic foam system components and manufacturing sequence

stretch forming: Plastic sheet-forming technique where thermoplastic sheet is stretched over a mold.
striae: Imperfections in plastic materials.
striation: A color variation and separation of colors resulting in a linear effect.
stringiness: An adhesive property resulting in the formation of filaments or threads when the adhesive transfer surfaces are separated.
structural adhesive: The bonding material required for transferring required loads between adherends in an environment.
structural bond: One that joins static or dynamic load-bearing parts of an assembly.
strux: A brand name describing CCA (cellular cellulose acetate) foam material used primarily to form structural stiffeners and cores in FRP structures.
styrene plastics: Group of plastics whose resins are derived from the polymerization of styrene or the copolymerization of styrene with various unsaturated compounds.
styrene-rubber plastics: Plastics that are composed of a minimum of 50% styrene plastic with the remainder rubber compounds.
sublaminate: Repeating orientations within a laminate (ply or laminae).
substrate: The surface to which various materials are applied or bonded.
sudden failure: Failure that could not be anticipated by prior examination or monitoring.
superfines: The portion of a powder (less than 10 microns) in powder metallurgy having particles smaller than a specified size.
superplastic forming (SPF): Strain rate sensitive metal forming process using materials with high elongation-to-failure properties.

surface mat: Very thin or fine mat, located at the surface of a laminate to ensure a smooth finish on the part.
surface preparation: Chemical and/or physical conditioning, of an adherent to make it suitable for adhesive joining, coating or other application.
surface resistance: The surface or electrical resistance between two electrodes in contact with a material surface is the ratio of the voltage applied to the electrodes to that portion of the current between them which flows through and along the surface layers.
surface resistivity: The electrical surface resistivity of a material is the ratio of potential gradient parallel to the current along its surface, to the current per unit width of surface.
surface tissue: A thin mat of fine fibers used primarily to produce a smooth surface on a reinforced plastic.
surface treatment: A chemical or additive treatment process which alters the surface characteristics of fibers and other materials improving bonding characteristics whether an additive or a chemical treatment.
swelling: Volumetric material increase resulting from temperature, moisture absorption or other condition.
symmetrical laminate: Composite laminate where the ply orientation has the same stacking sequence on either side of the laminate midplane.
syneresis: Contraction of a gel, observed by the separation of a liquid from the gel.
syntactic foam: Composites or lightweight materials composed of hollow microspheres or balloons and a resin system.
synthetic resin: Complex, substantially amorphous, organic semisolid or solid material (usually a mixture).

T

tab: An extension of composite or other material at each end of a tensile specimen to promote failure away from the grip area.

tab gate: Small removable tab approximately the same thickness as a molded item, usually located perpendicular to the item.

tack: Amount of stickiness for an adhesive or fiber reinforced resin system prepreg material. The property and ability of an adhesive to immediately bond to an adherend upon contact.

tack free: The stage at which a surface of coating, casting or paint is free from sticking to a material or other object it comes into contact with.

tack range: A time period during which an adhesive remains in the tack dry condition after being applied to an adherend, in a specified environment.

tack stage: A time period during which an adhesive film is tacky or sticky resisting removal.

tacky dry: A description of a coating or adhesive after the volatile constituents have evaporated or been absorbed.

Taguchi methods: The statistical design of experiments and analysis of variance methods in industrial design and production.

tangent modulus: Slope of the line at any predefined point on a static stress-strain curve, expressed in force per unit area per unit strain.

tape: Composite ribbon shaped material consisting of unidirectional continuous or discontinuous parallel fibers in a resin matrix aligned along the tape axis.

tape laying: A placement fabrication process in which the prepreg tape is laid side by side and/or overlapped forming a laminate structure.

taper plys: A blend of reinforcement or tapering-off plys placed in specific increments.

T-die: An inverted "T" center-fed slot, film extrusion die.

tear ply: The outer (peel) ply of material placed on a laminate that is sacrificial and when removed provides a rough, bondable surface.

tear resistance: Ability to resist a force along the edge of a material or test specimen.

telegraphing: Condition in a composite laminate where imperfections or patterns are transmitted to the surface.

temperature, curing: The use of temperature to cure an adhesive or other material.

template: Tooling aid shape, guide or pattern used for cutting and laying plies.

tenacity: Strength (ultimate tensile strength) of a filament of given size or yarn tenacity equals breaking strength (grams) divided by the denier.

tensile bar: A specimen (specified dimensions) used to measure tensile properties of a material and made by injection or compression molding.

tensile modulus: Ratio of a material's tension stress to strain over the range for which this valve is constant.

tensile strength: Maximum tensile load or force per unit area of original cross section, sustained by a specimen during a tension test.

tensile stress: Stress caused by forces directed away from the plane on which they act and applied force per unit of original crossectional area of specimen.

terpolymer: Polymeric system containing three monomers such as ABS (acrylonitrile, butadiene, styrene) terpolymer.

test: A critical trial (often involving stress) or examination of one or more properties or characteristics of a material, product, or set of observations.

test coupon: A portion, extension or part of a specimen or part used to analyze and determine the acceptability of a product.

test pattern: A mechanical or theoretical pattern that is used to inspect or test materials or products.

test program: The specifications or instructions controlling the test procedure.

test specification: The document that describes in detail the methods of conducting tests including if necessary, the criteria for assessing the result.

tex: A linear density unit equal to the mass or weight in grams of 1000 meters of fiber, filament, yarn, or other textile strand.

textile fibers: Filaments or fibers that can be processed into yarn or a fabric.

thermal conductivity: Ability of a material to conduct heat passing through a unit cube of the material in unit time.

thermal expansion coefficient: *See* coefficient of thermal expansion.

thermal expansion mismatch: A difference between the thermal expansion of two materials or components.

thermal expansion molding: A molding process in which elastomers are constrained within a rigid mold or frame and generate outward pressure by thermal expansion during the curing cycle.

thermal fatigue: The failure and fracture of a material or structure resulting from cyclic stresses as a result of temperature changes.

thermal load: A hygrothermal load component. A difference between the cure and operating temperature of a thermoset material can result in in-plane thermal loads for symmetric laminates; and both in-plane and flexural thermal loads for unsymmetric laminates.

thermal shock: Development of severe stresses following exposure to large temperature changes.

thermal stability: The ability of a material to remain stable when subjected to temperature change.

thermal stress cracking (TSC): Cracking and crazing of some thermoset and thermoplastic resin system resulting from over-exposure to elevated temperatures.

thermocouple: Two wires of dissimilar metals or alloys are connected (welded) to form a circuit. As the temperature varies a small electromotive force (EMF), or current, is measured and calibrated to various temperatures.

thermoelasticity: Rubber-like elasticity resulting from rigid plastic being exposed to increase of temperature.

thermoelectric power: Generating electric or magnetic fields in a solid as a result of change in temperature.

thermoforming: Forming a thermoplastic material, fibrous preform or other material by heating it to the point where it is soft enough to be formed over or into a mold contour and cooled.

thermogravimetric analysis (TGA): The procedure to study the mass of a material under various conditions of pressure and temperature.

thermoplastic: Plastic material capable of being repeatedly softened by application of heat and hardened by cooling.

thermoset: Plastic material which changes (cures, cross-links) into an infusible and insoluble material at room temperature or after the application of heat. (A chemical change).

thick molding compound (TMC): Three-dimensional random molding compound typically 6–50 mm thick.

thinner: A volatile liquid or solvent system added to an adhesive, coating or other material to reduce viscosity or the solids content of another substance.

thixotropic (thixotrophy): Condition of fluids whose apparent viscosity decreases with time to an asymptotic value under conditions of constant shear rate. Thixotropic fluids undergo a decrease in apparent viscosity by applying a shearing force such as stirring. If shear is removed, the material's apparent viscosity will increase back to or near its initial value at the onset of applying shear.

thread count: The number of threads or yarns per inch in either the (lengthwise) or fill (crosswise or weft) direction of woven fabrics.

thread, plug: A mold part used to shape an internal thread and which is unscrewed from the finished piece.

throwing: Term used to describe twisting and/or plying of strands in a yarn.

tie bars: Tooling components which provide structural rigidity to the clamping assembly of a press usually used to guide platens.

time, curing: Period of time during which a material, component or assembly is subjected to heat or pressure, or both, to cure the associated material.

time, drying: Period of time during which a coating or adhesive on an adherend or an assembly is allowed to dry with or without the application of heat or pressure, or both.

time, setting: The period of time during which a material, component or assembly is subjected to heat or pressure, or both, to set the associated material prior to curing.

toggle: A mechanical device that applies pressure through application of force on a knee joint and is used to close and exert pressure on tools, fixtures and molds in or out of a press.

tolerance: Exact allowance or maximum deviations from the standard dimensions, weight or stated environmental conditions.

tolerated stress: The stress under which the reliability characteristics of an item reach the set limit.

tooling: Molds or fixtures used for producing parts or assemblies.

tooling hole: An aid in the form of a hole (or holes) used to locate required positions in a fixture, mold or tool.

tooling resins: Thermoset resins systems (epoxy, polyester, silicone, etc.) used to fabricate cast or laminate master models, molds and tools.

tool life: The number of processing operations (cures or bonding operations) which can be successfully carried out on a particular tool.

tool side: Molded or bonded part side that is against the tool or mold surface.

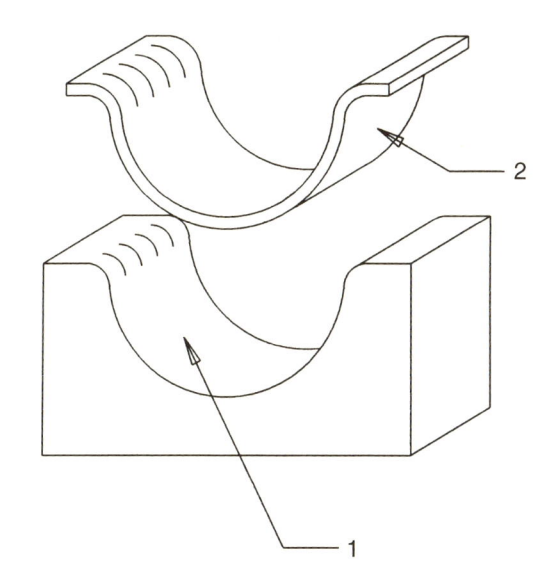

1–female molding tool surface
2–tool side of part

Tool side

torsion: Stress caused by twisting.

torsional rigidity (fibers): Ability of a fiber to resist twisting.

torsional strength: Stress resistance of materials being subjected to torque or twisting.

torsional stress: A shear stress caused by twisting applied to a transverse cross section.

toughness: A material property. The measure of energy required to break the material.

tow: Bundle of untwisted continuous filaments. Filaments are grouped into numbers designated by a "K" following a number (8K = 8,000 and 12K = 12,000 filaments).

traceability (international quality term): The ability to trace the history, application or location of an item or activity, or similar items or activities, by means of recorded identification.

tracer yarn (tracer): Strand of colored fiber distinct from the remainder of the fabric or roving package to identify warp from fill fibers and for quality checking. Also used to identify thickness of chopped fiber spray layups.

tracking: A condition where a high-voltage current source creates a conductive path across the surface of an insulating material by the formation of a carbonized path.

transducer: An electrical or mechanical force measuring device to measure pressure, compression and loads.

transfer molding: Process for molding thermosetting materials where the material is softened and pressure in a pot or transfer chamber is applied, then the material is forced by high pressure through runners, sprues, and gates into the closed mold the material cures and parts are ejected.

transfer molding pressure: Pressure applied across the area of the material pot or transfer chamber.

transformation: The variation of strength, stiffness, hygrothermal expansion, stress, strain and other parameters due to the coordinate transformation or, simply, the rotation of the reference coordinate axes.

transition temperature: The temperature at which the properties of a material change to or from a solid or rubber state.

transmittance: Light fraction that is transmitted through a substance.

transverse crack: Interfacial fiber and matric failure caused by excessive tensile stress applied transversely to the fibers in a unidirectional ply of a laminate.

transverse isotropy: A unidirectional composite material with symmetry that possesses an isotropic plane.

transverse strain: Linear strain plane perpendicular to the axis of a specimen.

trim line: Line that defines the borders or edge of a pattern, mold or part.

tumble winder: Filament winding machine with the mandrel mounted on a driven inclined axis with stationary roving strands. The tumbling mandrel describes a polar winding path.

tumbling: Finishing operation for articles and parts by which gates, flash, and other protrusions are removed, and surfaces are polished by rotating them in a shaped barrel (wet or dry) together with various types of aggregate, burnishing or polishing media.

turns per inch (TPI): The measure of twist produced in a yarn during its conversion from strand.

twill weave: A weave that repeats on three or more ends and picks and produces diagonal lines on the face of the cloth.

75

twist: In yarn or other textile strands the spiral turns about its axis per unit of length in a yarn or other textile strand.

twist, balanced: Twists of two or more strands that do not cause kinking or twisting on themselves when the yarn is held in the form of an open loop.

two-level mold: Single cavities stacked vertical above another thereby reducing clamping force.

U

ultimate elongation: The elongation of a tensile specimen at rupture.

ultimate strength: Maximum stress developed in a tension, compression or shear test specimen.

ultrasonic bonding: A process using ultrasonic frequency vibration energy and pressure which results in heat to form a joint.

ultrasonic cleaning: Cleaning that is done by passing high-frequency sound waves through a cleaning medium to cause agitation, loosening contamination from items to be cleaned.

ultrasonic insertion: Securing a metal insert into a plastic part by the application of vibratory mechanical pressure at ultrasonic frequencies. This action causes plastic to flow around the insert.

ultrasonic sealing: A film sealing method accomplished through the application of vibratory mechanical pressure at ultrasonic frequencies (20 to 40 kc) and electrical energy is converted to ultrasonic vibrations. This results in the generation of heat and the process is usually used to seal plastic films.

ultrasonic soldering: Soldering where molten fluxless solder is vibrated and heated at ultrasonic frequencies during making the joint.

ultrasonic testing: A nondestructive test for elastic sound-conductive materials used for locating voids and structural flaws within the material using an ultrasonic beam.

ultraviolet (UV) : Invisible radiations past the violet end of the spectrum of visible radiations. UV can be used to initiate cross-linking and chemical reactions. UV radiation will cause plastic degradation.

ultraviolet stabilizer (UVS): A chemical compound that, when mixed with a thermoplastic or thermoset resin system protects the system from UV degradation.

unbond: Area within a bonded interface between two adherends in which the intended bonding action failed to take place or an area deliberately masked from bonding, as in a test specimen. *See also* debond.

unconditional test: A test without limitations or restrictions on test mode, test time, etc.

undercure: A condition in which insufficient time and/or temperature was allowed in curing, resulting in inadequate hardening of the material.

undercut: A indentation, reverse or negative draft on a part in a mold.

underwriters laboratory symbol (UL): A mark or logo showing that a product has been recognized (accepted) by Underwriters Laboratories, Inc.

uniaxial load: A condition whereby a material is stressed in only one direction along the axis or centerline of component parts.

unidirectional: Reinforcement fibers oriented in the same direction.

unidirectional composite: A composite laminate having all reinforcing fibers of all plies oriented in the same direction.

unidirectional laminate: Same as unidirectional composite.

unidirectional roving: Parallel heavy highly directional strength roving with light rovings at right angles to them.

unsaturated compounds: Any compound with more than one bond between two adjacent atoms, usually carbon atoms, and capable of adding other atoms at that point to reduce it to a single bond.

unsaturation: The presence of double (or triple) bonds in a compound, which may take part in addition polymerization or cross-linking.

unsymmetric laminate: A laminate having an arbitrary stacking sequence without midplane symmetry.

up time: The period of time during which an item is in a condition to perform its required function.

urea plastics: Group of plastics whose resins are derived from the condensation of urea and aldehydes.

urethane plastics: Group of plastics composed of resins derived from the condensation of organic isocyanates with compounds containing hydroxyl groups.

useful life: The period from a stated time, during which, under stated conditions, an item has an acceptable failure rate, or until an unrepairable failure occurs.

V

vacuum: A pressure below atmospheric pressure and normally measured in "inches" of mercury which refer to the height (inches) of mercury in a tube supported by the outside pressure differential.

vacuum assisted resin injection (VARI): The process of loading fibrous reinforcement into a mold, then injecting resin to flow into the mold by the action of gravity or pressure assisted by evacuation of the mold followed by curing the resin before opening the mold. Also referred to as Resin Transfer Molding (RTM).

vacuum bag: An impervious flexible disposable film or reusable elastomer sheet which when sealed to the mold surface along the entire edge contains the laminate or part under the action of vacuum pressure.

vacuum bagging: Curing a laminate or bonded assembly in an evacuated bag to allow the application of additional consolidation pressure.

vacuum bag molding: A reinforced plastic molding process in which a sheet of elastomeric material or flexible transparent film material is placed over the layup on the mold surface and sealed along the entire edge. A vacuum is applied between the bag and the layup. The entrapped air is mechanically worked out of the layup under the bag and removed by the vacuum. The part is cured.

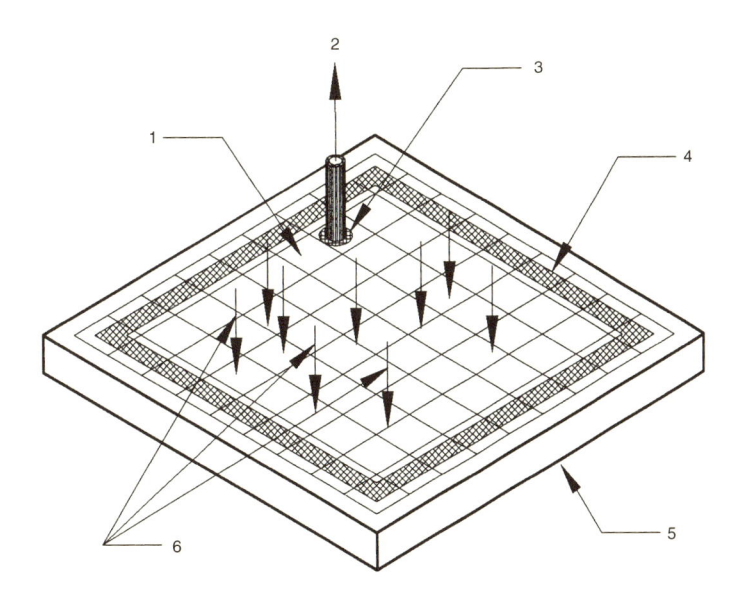

1–vacuum bag outside sealed to the sealant, breather is beneath the vacuum bag
2–air evacuated from beneath the bag
3–vacuum port sealed to vacuum bag
4–sealant tape applied to master mold or tool
5–master mold or tool
6–atmospheric pressure uniformly pressing down on the vacuum bag and whatever is beneath it

Vacuum bag

vacuum deposition: The application of a thin material coating onto a base material by evaporation techniques in a vacuum.

vacuum forming: Fabrication process in which fibrous, thermoplastic and thermoset plastic sheet materials are transformed into three dimensional shapes by inducing flow; accomplished by reducing the air pressure on one side of the sheet, applying vacuum and heat.

vacuum hot pressing (VHP): A processing method for forming powders and other materials at low atmospheric pressures, elevated temperatures and consolidation pressures.

vacuum impregnation: The process of loading reinforcement fibers into a mold then causing resin to flow into the mold by applying a vacuum, followed by curing the resin before opening the mold.

vacuum metalizing: A metal coating process in which surfaces are coated with a thin metal coating using a vapor of metal that has been evaporated and deposited under vacuum.

vacuum table: An impervious surface, plate, film or other material, which holds the laminate or part under vacuum bag pressure during curing.

variable: A characteristic that is appraised in terms of values on a continuous scale.

variance: The square of the average value of deviation from a mean value.

veil: An ultrathin fibrous glass or organic fiber mat material similar to a surface mat, and used as the surface layer in reinforced plastics.

vendor inspection lot (material): All of the same type material processed during a specified period of time using similar procedures and conditions that are offered for inspection at one time.

vent: A shallow channel, opening or hole in a mold that allows air, gas or excess material to exit while molding material enters.

venting: In autoclave processing of a laminated part or bonded assembly, it is venting the vacuum bag to the atmosphere after closing off the vacuum.

vermiculite: A granular mica-like material used as a filler with resin, resulting in high compression strength.

vibroforming: The use of vibrating rollers during consolidation in contact molding.

vicat softening temperature: Heat distortion or heat deformation temperature of a plastic material.

vinyl acetate plastics: Group of plastics composed of resins derived from the polymerization of vinyl acetate with other saturated compounds.

vinyl alcohol plastics: Group of plastics composed of resins derived from the hydrolysis of polyvinyl esters or copolymers of vinyl esters.

vinyl chloride plastics: Group of plastics whose resins are derived from the polymerization of vinyl chloride and other unsaturated compounds.

vinyl esters: Thermosetting resin systems containing esters of acrylic and/or methacrylic acids, many of which have been made from epoxy resin.

vinyl plastics: Group of plastics composed of resins derived from vinyl monomers, excluding those that are covered by other classification (i.e., acrylics and styrene plastics). Examples include PVC, polyvinyl acetate, polyvinyl butyral, and various copolymers of vinyl monomers with unsaturated compounds.

vinylidene plastics: Group known as saran plastics.

viscoelasticity: A property involving a combination of elastic and viscous behavior in the application of which a material is considered to combine the features of a perfectly elastic solid and a perfect fluid. Phenomenon of time-dependent, in addition to elastic, deformation (for recovery) in response to load.

viscometry: The measurement of viscosity.

viscosity: The material's fluidity property of resistance to flow exhibited within the body of a material, expressed in terms of relationship between applied shearing stress and resulting rate of strain in shear.

visual examination: Qualitative inspection and observation of physical characteristics using the unaided eye or specified magnification.

void: A pocket, empty space; bubble occurring within the thickness of a plastic or fibrous laminate.

void content: The percentage of voids in a laminate.

volatile: Capable of continually releasing vapors in air.

volatile content: Percentage of volatiles that are driven off as a vapor from a material.

volatile loss: Vaporization weight loss.

volume fraction: A portion or fraction of the total volume.

volume percentage: The percentage of one component relative to the whole body by volume.

vulcanization: Chemical reaction in which rubber or plastic physical properties are altered by reaction with other suitable agents.

W

wafer: Additional reinforcement for openings.

wall stress: Stress (usually in filament wound pressure vessels) calculated using the load and the entire laminate cross sectional area.

warp: Yarn or group of yarns running lengthwise and parallel to the selvage in a woven fabric. Also a change in dimension of a plastic part from its original molded shape. The construction of cloth or fabric consists of interweaving warp yarns (along the length of the fabric in the loom) and filling yarns (across the fabric in the loom).

warpage: Dimensional distortion. *See also* warp.

warp insertion weft knit fabric: Weft knitted fabric into which uncrimped warp fibers are inserted within the weft knitted loops.

warp knitting: Method of making a fabric by normal knitting in which the loops made by each warp thread are formed substantially along the length of the fabric.

washout core: Material able to be removed without disassembling a model or pattern formed around it.

water absorption: Ability of a material to absorb water from an environment.

water break: A discontinuous appearance of a water film on any surface.

water jet: High pressure stream of water emitted from a nozzle and useful for cutting organic composites.

water-white: Having the appearance or clarity of water.

waywind: Number of wraps or turns that an end or ends make from one side of the filament wound package back to the same side.

wear-out failure: Failure whose probability of occurrence increases with the passage of time and which occurs as a result of failure mechanisms that are characteristic of the population.

wear-out failure period: That possible period during which the failures occur at an approximately uniform rate.

weathering: Exposure of materials to the outdoor environmental conditions of high and low temperatures, high and low relative humidities, and ultraviolet radiant energy, with or without direct water spray.

weathering, artificial: Exposure of plastics to accelerated laboratory produced environmental conditions.

weatherometer: Instrument for subjecting articles to accelerated weathering conditions.

weave: The pattern (usually assigned a style number) by which a fabric is formed from interlacing yarns. A plain weave has the wrap and fill fibers alternating to make both fabric faces identical. A satin weave pattern produces a satin appearance.

weave exposure: Material surface condition where unbroken fibers of woven cloth are not completely covered by the resin system.

web: A paper, textile fabric, or thin metal continuous length sheet in roll form.

weeping: A slow internal leakage of a material from within to the surface.

weft: Transverse fibers or threads in a woven fabric running perpendicular to the warp (shuttle direction).

weft insertion warp knit fabric: Warp knitted fabric into which uncrimped weft fibers are inserted within the warp knitted loops.

weft knitting: A method of making a fabric by normal knitting in which the loops made by each weft thread are formed substantially across the width of the fabric.

weight fraction: The proportion of one component relative to the whole body by weight.

weight percentage: The percentage of one component relative to the whole body by weight.

welding: Joining thermoplastic pieces by one of several heat-softening processes (butt fusion, spin welding, ultrasonic, and hot gas, etc.).

weld lines: Molding marks visible on a finished part made by the meeting of two resin flow fronts.

wet: Having absorbed moisture and located near the exposed surface.

wet flexural strength (WFS): Flexural strength after

wet layup

water immersion, usually after boiling a test specimen for 2 hr. in water.

wet layup: A reinforced plastic molding method where you apply the resin system as the reinforcement is put in place.

wetout: Saturation or soaking of porous materials such as reinforcement strands and filaments.

wetout rate: Time required for a plastic resin system to fill the interstices of reinforcement material and wet the fiber surfaces.

wet strength: Strength of an organic matrix composite after it is at a defined percentage of absorbed moisture, less than saturation.

wetting: Absorption or spreading of a liquid into or along a surface.

wetting agent: An agent on the surface of a material that produces wetting by decreasing cohesion within the liquid and causing intimate contact to the surface.

wet winding: A filament winding process in which the fiber reinforcement is coated with resin immediately prior to wrapping on the mandrel.

whisker: A very short single crystal reinforcement fiber form.

wicking: Capillary absorption of a liquid into or along reinforcement fibers.

wind angle: The angle between the axis of rotation and filament wound fibers.

winding pattern: The regular recurring filament winding path pattern.

winding tension: In filament winding, the amount of tension on the reinforcement during the winding process.

window: A total range of evaluation parameter values.

wire: A metal filament.

witness: The arrival of resin at the exit port of a mold.

wood veneer: Thin sheet of wood.

work hardening: Hardening and strengthening of a material by the strain energy absorbed from prior deformation.

work life: Period during which a catalized thermosetting material compound remains usable for its intended application.

woven fabric: Interlaced yarns, fibers or filaments forming a planar mesh pattern (plain, twill, satin).

woven roving: Heavy fabric made by the weaving of roving or bundles of yarn.

wrinkle: A laminated plastic material surface imperfection or crease in one or more outer layers of a laminate.

X

x-axis: In a composite laminate is the axis in the plane perpendicular to the y-axis. Also is zero degrees reference for designating other angles of the lamina.

XY-plane: In a composite laminate, plane of reference parallel to the plane of the laminate.

Y

yarn: An assemblage of natural or made twisted fibers or strands that form a continuous yarn to be used in weaving or interweaving into textile materials. *See also* filament.

yarn, plied: Two or more yarns twisted together in one operation.

yarn, warp: Identifies where yarn is used within a fabric, i.e., in warp direction.

yarn count: Specific length per unit mass.

y-axis: In a composite laminate, the axis in the laminate plane perpendicular to the x-axis.

yield: Amount of material equivalent to unit weight.

yield point: The first point where stress in a material, less than the maximum attainable stress occurs. The point where an increase in strain occurs without an increase in stress. Materials that exhibit yielding have a yield point.

yield strength: A point of lowest stress where a material undergoes plastic deformation. Below this stress point, the material is elastic and above, viscous.

yield value: The measure of normal or shear stress at which a marked increase in deformation occurs without an increase in load.

Young's modulus: A ratio of normal stress to corresponding axial strain for tensile or compressive stresses less than the proportional limit of the material. *See also* modulus of elasticity.

Z

z-axis: In a composite laminate, the reference axis normal to the plane of the laminate.

zero bleed: Describes a prepreg made with the final amount of resin desired in the part. No resin has to be removed during cure. Also a manufacturing process that eliminates resin loss during cure.